KEY NOTES ON AGRICULTURE BOTANY

For Ready Reference to the

STUDENTS, TEACHERS, RESEARCHERS & ASPIRANTS OF COMPETITIVE EXAMINATIONS

THE EDITORS

Dr. U.D. Chavan obtained his M.Sc. (Agri. in Biochemistry) degree from Mahatma Phule Krishi Vidyapeeth, Rahuri. He received his Ph.D. degree in Food Science from Memorial University of Newfoundland St. John's Canada in 1999. He has done International Training on "Global Nutrition 2002" at Uppsala University Uppasala, Sweden in 2002. Dr. Chavan worked as Senior Research Assistant in the Department of Biochemistry & Food Science and Technology at MPKV Rahuri from 1988 to 2000. During his Ph.D., he worked as Technician/Research Associate at Atlantic Cool Climate Crop Research Center and Agriculture and Agri-Food Canada. He received D.Sc. degree in 2006 from USA.

Dr. Chavan is presently working as a Senior Cereal Food Technologist in the Department of Food Science & Technology at Mahatma Phule Krishi Vidyapeeth, Rahuri.

Dr. J.V. Patil obtained his M.Sc. (Agri.) from, MPKV, Rahuri. He completed his course work for Ph.D. at CCSHAU, Hisar and research at MPKV, Rahuri in 1992. He rendered his research and teaching services at MPKV Rahuri as Geneticist, Associate Professor, Plant Breeder and Professor of Genetics & Plant Breeding and Head, Genetics and Plant Breeding Department, MPKV, Rahuri. He also delivered many administrative responsibilities in the University. Dr. Patil joined as the Director, Directorate of Sorghum Research, Hyderabad in August 2010.

THE CONTRIBUTORS

Dr. K.M. Pol is a Professor in the Department of Agriculture Botany at Mahatma Phule Krishi Vidyapeeth, Rahuri.

Dr. A.M. Dethe is an Assistant Professor in the Department of Agriculture Botany at Marathwada Agricultural University, Parbhani.

Dr. N.S. Kute is an Associate Professor in the Department of Agriculture Botany at Mahatma Phule Krishi Vidyapeeth, Rahuri.

KEY NOTES ON AGRICULTURE BOTANY

For Ready Reference to the

STUDENTS, TEACHERS, RESEARCHERS & ASPIRANTS OF COMPETITIVE EXAMINATIONS

Editors

U.D. CHAVAN
&
J.V. PATIL

Contributors

K.M. POL
A.M. DETHE
N.S. KUTE

2015

Daya Publishing House®

A Division of

Astral International (P) Ltd

New Delhi 110 002

Published by   : **Daya Publishing House®**
                           A Division of
                           Astral International Pvt. Ltd.
                           – ISO 9001:2008 Certified Company –
                           4760-61/23, Ansari Road, Darya Ganj
                           New Delhi-110 002
                           Ph. 011-43549197, 23278134
                           E-mail: info@astralint.com
                           Website: www.astralint.com

Laser Typesetting   : **Twinkle Graphics, Delhi**

Printed at   : **Thomson Press India Limited**

PRINTED IN INDIA

PREFACE

India is an agricultural country. The Indian economy is basically agarian. Inspite of economic and industrialization, agriculture is the backbone of the Indian economy. As Mahatma Gandhi said "India's lives in villages and agriculture is the soul of Indian economy". Agriculture is a vast subject and encompasses at least 20 major and minor subjects in it. New developments have lead to entirely a new face of agriculture. Study of agriculture has always been intrigued with a mosaic of interwove concepts, subjects, facts and figures. There are number of books and large literature on Agriculture Botany but the Key Notes type of book have not been compiled in a readable manner.

The present book *"Key Notes on Agriculture Botany"* has been designed to fulfill this long felt need of students, teachers, researchers and aspirants of competitive examinations. It is designed in such a way that give rapid, easy access to the core materials in a short format which facilitates easily learning and rapid revision. The book carries fundamentals of Agriculture Botany. There are 8 chapters elaborating Discoveries, Abbreviations, English/Scientific Names of Fish, Poultry and others, Terminology, Reasoning/Short Notes, Differences/comparing, Concepts, Theories and Laws as well as references also included. The most recent information is provided along with a detailed list of references for further reading.

Hope this book would be highly useful for graduate and post-graduate students of agriculture, teachers and researchers. This book will also useful for the aspirants of various competitive examinations such as Agricultural Research Service (ARS), ICAR- National Eligibility Test (NET), State Eligibility Test (SET), Junior Research Fellowship (JRF), Senior Research Fellowship (SRF), Civil Services, Allied Agricultural Examinations and Extension Workers for reference and easy answers of many complicated questions. Thus it is expected that this book will adequately meet the need of wider circle of students and readers for preparing their professional career.

We acknowledge the references that are used in this manuscript. Authors are also thankful to all scientists and friends who have helped directly or indirectly while preparing this manuscript. The editors of grateful to all the contributors for their cooperation, support and timely submission of their manuscripts for bringing out this publication. We would have like to acknowledge the patience and support of our families whilst we have spent many hours with drafts of manuscripts rather than with them. Lastly, our sincere thanks to publisher Astral International Pvt. Ltd., New Delhi who provides an opportunity to publish this book.

To all readers we extend an invitation to report that no doubts have escaped our attention and to offer suggestion for improvements that can be incorporated in future editions.

U.D.Chavan and J.V. Patil

Editors

CONTENTS

1

DISCOVERIES

Scientist	Year	Discovery
Blackman, F.F.	-	Dark reaction (chemical reactions)
Emerson	-	Red drop
Altmann's theory	1886-90,94	Granule theory
Antony van Leeuwenhock and Rebert Hooke	1632-1723 1635-1703	Termed Cell
Stephen Hales	1727	Root pressure theory
Carl-Nageli anthocyanins	1867	Impermeability of the protoplasm to red, purple or blue
Trauble	1867	Haftdruck or retention pressure theory
Schimper	1883	Green plastids as chloroplast
Godlewski	1884	Relay pump theory
Camille Golgi	1891	Golgibodies
Neljubow	1901	Discovery of strange phenomenon induced by a gas
Biffen	1902	Reported that resistance to strip rust in wheat was due to a single recessive gene was the first explanation of the nature of inheritance to disease reaction.
Schroeter & Kircher	1902	Autoecology, synecology – Term coined
Lutz	1907	He described the first variation in chromosome number, *i.e.*, heteroploidy, discovered in an experimental population was the gigas mutant in *Oenothera*.
Warming	1909	Plants classified on the basis of (Water relation) adaptation
Barrus	1911	Showed that different isolates of microorganism (bean anthracnose fungus, *Colletotrichum lindemuthianum*) differed in their ability to attack different varieties of the same host species; this finding is the basis for physiological races or pathotype.
Rubland, W., Kuster, and Hoffman	1912-25	Cell membrane act as molecular sieves or as ultrafilter.
Hays and Garbar	1919	Were the first to advocate commercial utilization of synthetic varieties?
Belling	1920	Demonstrated the globe mutant of *'Datura stramonium'* developed by Blakeslee to be a trisomic.
Loftfield	1921	Diurnal changes in starch/sugar balance in the guard cell
Mc Hargue	1922	Manganese

Contd...

Scientist	Year	Discovery
Bose J.C.	1923	Vital theory for the ascent of sap
Warington	1923	Boron
Gorter and Grendel	1925	Lipid in a monolayer
Curtis	1926	Transpiration as the necessary evil
Sommer and Lipman	1926	Zinc
Muller	1927	He showed for first time that mutations could be induced in Drospophila through treatment with 'X-rays'
Keller	1930	Water is absorbed by the xylem by electroosmotic mechanism
Lipman and Mac Kinney	1931	Copper
Crocker *et. al.*	1932	Investigation of ethylene physiology
Gane	1934	Ientification of the gas as ethylene
Danielli and Davson	1935	Bimolecular leaflet of lipid
Davson-Danielli	1935	Proteins are in the globular (alpha) configuration
Robertson	1935	Proteins are in the globular (beta) configuration
Balkeslee and Avery	1937	Showed that chromosome doubling can be induced in plants by treatment with colchicine.
Meyer	1938	The term DPD was introduced
Arnon and Stout	1939	Molybdenum
Sprague and Tatum	1942	Coined the words 'general combining ability (GCA)' and specific combining ability (SCA)
Roberston, and Turner Lundegardh	1945-55	Oxidase inhibitors, azides and cyanides strongly inhibited salt uptake
Kidd and West	1945	Involvement of ethylene in fruit ripening
Beadle and Tatum	1946	One gene one protein
Wilson	1948	Light-induced opening of tomata
Clavin	1948	TCA cycle
Garbrielsen	1948	Photosynthesis rate in different light
Williamson	1950	Ethylene formation of infiltrated plants
Horecker *et al.* Racker	1951-54	Elucidated pentose phosphate pathway reactions
Painter	1951	Published book on breeding for insect resistance 'Insect Resistance in Crop Plants'. The types of insect resistance *i.e.* (1) Non preference, (2) antibiosis, (3) tolerance were recognizd by Painter and 4th category avoidance included subsequently.
Robinson and Brown	1953	Ribosomes in plant cells
Palade	1953	Ribosomes in animal cells
Dalton and Ferma	1953	Structure of Golgibodies
Watson J.D. and Crick F.H.C	1953	Structure of DNA

Contd...

Scientist	Year	Discovery
Frey-Wyssling Wolken Hodge *et. al,* Calvin	1953-53 1955-59	Molecular model of the chloroplast
Broyer and others	1954	Chlorine
Addicott *et al.*	1955	Auxin gradient hypothesis
Osborne	1955	Auxin senescence factor hypothesis
Gaur and Leopold	1956	Auxin concentration hypothesis
Griffing	1956	Gave procedures for estimation of GCA and SCA effects and genetic components of variation.
Flor	1956	Working with flax rust fungus, *Melampsora lini,* proposed classical gene for gene hypothesis, which states that for each gene conditioning resistance in host, there is a corresponding gene in parasite conditioning pathogenicity.
Heath and Orchard	1957	Mid day time stomatal closure is more
Kempthorne	1957	The concept of partial diallel and Line x tester technique developed by him.
Hall *et. al.*	1959	Metabolic fate of ethylene ^{14}C
Meigh	1959	Gas chromatography of ethylene
Burg and Thimann	1959	Beginning of research of ethylene biosynthesis
Jensen	1960	Pectic substances in cell
Hendler	1962	Scheme for protein synthesis
Weier and Thomson	1962	Grana
Burg and Burg	1962	Fruit ripening hormone
Dainty and Gutknecht	1963 1967	Pore may be hydrophilic regions in the lipid layer
James and Richens	1963	Ribosomes for protein synthesis in nucleus
Brinstiel and Hyde	1963	Ribosomes for protein synthesis in nucleolus
Van der Plank	1963	Proposed two main types of resistance, *viz.;* vertical and horizontal resistance.
Das and Mukherjee	1964	Ribosomes for protein synthesis in mitochondria
Osgood et al	1964	Chromosomes consists of coils of 16 double strand DNA like a rope
Abeles & Rubinstein	1964	Auxin induced ethylene formation
Kavanau	1965	Holes in the plasma form and reform
Weier *et al.*	1965	Three-dimensional picture of the chloroplast lamellae
Spencer	1965	Ribosomes for protein synthesis in chloroplast
Lieberman *et. al.*	1965	Discovery of methionine as a precursor
Shimokawa and Kasai	1965	Liquid scintillation counting method of ethylene ^{14}C
Lenard Singer	1966	Proposed lipid-protein-lipid structures
Suzama and Bonner Breied-Enbach *et. al.*	1966-67	Discovered DNA and RNA within mitochondria

Contd...

Scientist	Year	Discovery
Haekel	1966	Plastids
Abeles	1966	Ethylene-induced abscission
Hatch, M.D. and Slack, C.R.	1967	Hatch-Slack cycle in C_4 plant and only in *Triplex hastata* temperate species.
Burg and Burg	1967	CO_2 a competetitive inhibitor of ethylene action
Lehninger	1968	Membrane model
Hadziyer *et al.*	1968	Presence of specific DNA and RNA in chloroplast
Cooke and Randall	1968	Discovery of ethephone-an ethylene releasing compound
Abeles	1968	Phytyogerontology
Chen and Wildman	1970	The chloroplast DNA synthesizes protein
Dupraw	1970	The nuclonema consists the template for the 16 S and 28 S ribosomal RNAs.
Shimokawa and Kasai	1970	Preparation of an enzyme catalyzed ethylene formation from acrylate
Hallauer and Eberhart	1970	Proposed the Full-sib reciprocal recurrent selection method of population improvement.
Kirk	1971	Model of thylakoid membrane
Owens *et. al.*	1971	Inhibition of ethylene formation with enol ether amino acid analogs
Vendrell *et. al.*	1971	Autoinhibition of ethylene formation
Harlan and Dewet	1971	Gave the concept of 'gene pool' for effective and judicious use of germplasm.
Singer & Nicholson	1972	Cell permeability due to lipid and protein interactions
Green and Capaldi	1974	Fluid-mosaic model
Raschke	1975	Stomatal movement
Racker	1976	Model of the membrane with proteins
King, T.E.	1977	Location of enzymes in the membrane
Adams and Yang	1977	Establishment of ethylene biosynthesis from methionine
Lurssen *et.al.* Adams & Yang	1979- 1979	Discovery of ACC as a precursor
Konze and Kende	1979	
M.S. Swaminathan	1980	The Director General, ICAR, New Delhi associated with development of semidwarf wheat and rice varieties in India, resulted in Green Revolution. Father of GREEN REVOLUTION in India.
Comstock and Robinson	1948 1952	The concept of biparental mating was developed by them. Plants are randomly selected in F2 or subsequent generation of cross between two purelines having contrasting perfor-mance. The selected plants are crossed according to definite scheme.
Harlan	1992	Coined the term 'micro centers' with in the centers of diversity, the crop show enormous diversity in very small regions (100-500 kms across).

2

ABBREVIATIONS

Abbreviation	Full Form
AASCO	Association of American Seed Control Officials
ABA	Abscisic acid
AFC	Agricultural Finance Corportaion
AICCIP	All India Co-ordinated Crop Improvement Project.
AICRIP	All India Coordinated Research Improvement Project
AICRP	All India Co-ordinated Research Project
AOAC	Association of Official Analytical Chemists
AOSA	Association of Official Seed Analysts
AOSCA	Association of Official Seed Certifiying Agencies
APC	Agro-processing Center
APS	Ammonium Persulphate
ARARI	Aegean Regional Agricultural Research Institute
ARC	Agricultural Refinance Corportation
AS	Ammonium Sulphate
ASSINSEL	International Association for the Protection of Plant Breeder Rights
ASTA	American Seed Trade Association
AUDPC	Area Under Disease Progress Curve
AVRDC	Asian Vegetable Research and Development Center
BA	Benxyladenine
BAPNA	Benzeyl-DL-arginine-p-nitroanilide
BASIC	Beginners All Purpose Symbiotic Instruction Code
BIOS	Basic Input Output System
BIS	Bureau of Indian Standards, New Delhi

Abbreviation	Full Form
Bit	Binary Digit (1 Byte = 8 Bits)
BP	Between Paper
BSI	Botanical Survey of India
CABI	Center for Agriculture and Biosciences International
CAC	Codex Allmentarius Commission
CAM	Crassulaceae Acid Metabolism
CAN	Calcium Ammonium Nitrate
CARI	Central Agricultural Research Institute for Andaman and Nicobar Group of Island
CAZRI	Central Arid Zone Research Institute, Jodhpur
CAZRI	Central Arid Zone Research Institute
CC	Core complex
CCCP	Carboxyl cyanamide m-chlorophenyl hydrozone
CEC	Cation Exchange Capacity
CFC	Fluro Floro Carbon
CG	Canadian Genebank
CGIAR	The Consultative Group on International Agricultural Research
CGPRT	Center for Research and Development of Coarse Grains, Pulses, Roots and Tuber Crops
CGR	Crop growth Rate
CHAs	Chemical Hybridising Agents
CIAE	Central Institute of Agricultural Engineering
CIAT	International Center for Tropical Agriculture
CICR	Central Institute for Cotton Research
CIMMYT	International Maize and Wheat Improvement Center
CIP	International Potato Center
CK	Cytokinins
CMEA	Council for Mutual Economic Assistance
CN	Carbon Nitrogen
CPCB	Central Pollution Control Board

Abbreviation	*Full Form*
CPCRI	Central Plantation Crops Research Institute
CPP	Copalyl pyrophosphate
CPRI	Central Potato Research Institute
CRI	Critical Root Initiation Stage
CRIDA	Central Research Institute for Dryland Agriculture
CRRI	Central Rice Research Institute
CSC	Central Seed Committee
CSCB	Central Seed Certification Board
CSSRI	Central Soil Salinity Research Institute
CSTL	Central Seed Testing Laboratory
CSWCRTI	Central Soil and Water Conservation Research and Training Institute
CTCRI	Central Tuber Crops Research Institute
CTRI	Central Tobacco Research Institute
CTRL	Cotton Technological Research Laboratory
CV	Coefficient of Variation
DAP	Di-Ammonium Phosphate
DCA	2, 4-dichlororanizole
DIP	Destructive Insect Pest
DNP	Dinitrophenol
DNSA	3-5, Dinitro salicylic acid
DPD	Diffusion Pressure Deficit
DSCO	Divisional Sub-Commanding Officer
EC	Electrical conductivity
EEC	European Economic Comunity
EIL	Economic Injury Level
EMP	Error Mean Sum of Product
EPA	European Production Agency
EPPO	Europian and Mediterranean Plant Protection Organization
ER	Endoplasmic reticulum

Abbreviation	Full Form
ETL	Economic Threshold Level
EUCARPIA	Europian Association for Research on Plant Breeding
FAI	Fertilizer Association of India
FAO	Food and Agricultural Organization
FCI	Food Carporation of India
FFHC	Fredom From Hunger Campaign
FIS	Federation of International Seed Trade
FYM	Farm Yard Manure
GA	Gibberellic Acid
GA	Genetic Advance
GATT	General Agreement on Tariffs and Trade
GCA	General Combining ability
GCV	Genotypic Coefficient of Variation
GDP	Gross Domestic Products
GGPP	Geranylgeranyl pyrophosphate
GIS	Geographical Information System
GMP	Genotypic Mean Sum of Products
GOI	Government of India
GSI	Geological Survey of India
HACCP	Hazard Analysis Critical Control Point
HIR	High irradiance reactions/responses
HRGP	Hydroxy proline rich glycoproteins
HSS	Herbage Seed Scheme
HYV	High Yielding Variety
IAA	Indole-3-acetic acid
IAPI	International Association of Plant Taxonomy
IAPSC	Inter African Phyto-Sanitary Commission
IARI	Indian Agricultural Research Institute
IASRI	Indian Agricultural Statistics Institute, New Delhi.
IASRI	Indian Agricultural Statistical Research Insittute

Abbreviation	Full Form
IBP	International Biological Programme
IBPGR	International Board of Plant Genetic Resources
IBRD	International Bank for Reconstruction and Development
ICAR	Indian Council of Agricultural Research
ICARDA	International Center for Agricultural Research in Dry Areas
ICGEB	International Center for Genetic Engineering Biotechnology
ICP	International Center for Potato
ICRISAT	International Crop Research Institute for the Semi-Arid Tropics
ICTA	International Center for Tropical Agriculture
IGFRI	Indian Grassland and Forage Research Institute
ILCA	International Livestock Center for Africa
IIHR	Indian Institute of Horticulture Research
IISB	Indian Institute of Sugarcane Breeding
IISR	Indian Institute of Spices Research
IISRI	Indian Institute of Sugarcane Research
IITA	Internatinal Institute for Tropical Agriculture
ILRI	Indian Lac Research Institute
IPB	Institute of Plant Breeding
IRRI	International Rice Research Institute
ISO	International Standard Organization
ISSR	Inter Simple Sequence Repeats
ISST	Indian Society of Seed Technology
ISTA	International Seed Testing Association
JARI	Jute Agricultural Research Institute
JTRL	Jute Technological Research Laboratory
Km	Michaelis-Menten constant
LAI	Leaf area index
LDP	Long day plants
LHC	Light harvesting complex
LSDP	Long-short-day plants

Abbreviation	Full Form
MA	Milli Ampere
MCPA	2-Methyl-4-chloro-phenoxyacetic acid.
MH	Maleic hydrazide (1, 2-dihydro-3, 6-pyridazinedione)
MLD	Minimum lethal dose
MM	Milli Molar
MOP	Muriate of Potash
MPC	Marginal Propensity to Consume
MSB	Menadione sodium bisulfite
MSS	Mean Sum of Squares
MSSC	Maharashtra State Seeds Corporation
MVA	Mevalonic acid
NAARM	National Acadamy of Agricultural Research Management
NABARD	National Bank for Agriculture and Rural Development
NARP	National Agricultural Research Project
NBPGR	National Bureau of Plant Genetic Resources
NCA	National Commission on Agriculture
NCDC	National Cooperative Development Corporation
NEPPC	Near East Plant Protection Commission
NG	Nordic Genebank
NIAS	National Institute of Agricultural Sciences
NSC	National Seeds Corportation
NSDC	National Seed Development Council
NSP	National Seeds Programme
NSP	National Seed Project
NSRS	National Seeds Research Station
NSSL	National Seed Storage Laboratory
OCP	Other Crop Plants
OECD	Organization of Economic Cooperation and Development
OEEC	Organization for Europen Economic Community
OP	Osmotic pressure

Abbreviation	Full Form
OW	Obectionable Weeds
PAGE	Polyacrylamide gel electrophoresis
PAR	Photosynthetically active radiation
PCD	Programmed cell death
PCO	Photorespiratory carbon oxidation
PCV	Phenotypic Coefficient of Variation
PDI	Percent Disease Index
PEP	Phosphoenol Pyruvate
PEQ	Post Entry Quarantine
PER	Preotein Efficiency Ratio
PGA	Phosphoglyceric acid
PGI	Plant Genetic Institute
PGRC	Plant Genetic Resources Center
pH	Negative logarithm of the hydrogen ion concentration
PLS	Pure Live Seed
PMA	Phenylmercuric acetate
PUFA	Polyunsaturated Fatty Acids
PWP	Permanent wilting point
R.Q.	Respiratory ratio or Quotient
RAPD	Random Amplified Polymorphic DNA
RC	Reaction center
RCNEHR	ICAR Research complex for North-Eastern hill region
REM	Radiation Effect on Man
RGR	Relative growth rate
Rm	Relative migration
RO	Reverse Osmosis
rpm	Revolution Per Minute
RPP	Reductive pentose phosphate
RuBP	Ribulose biphosphate
SAAC	Commercial Seed Analysts Association of Canada

Abbreviation	Full Form
SAM	S-adenosyl methionine
SAR	Systematic Acquired Resistance
SARCCUS	Southern African Regional Commission for the Conservation and Utilization of Soil
SBD	Seed Borne Diseases
SBI	Sugarcane Breeding Institute
SCA	Specific Combining Ability
SCO	Sub-Commanding Officer
SCST	Society of Commercial Seed Technologists
SDP	Short day plants
SDS	Sodium-dodecyl Sulphate
SEAPPPC	Plant Protection Committee for the South East Asian and Pacific Region
SFCI	State Farm Corporation of India
SI	Stomatal index
SIDA	Swedish International Development Authority
SIDP	Seed Improvement and Development Programme
SLDP	Short-long-day plants
SRT	Seed Review Team
SSC	State Seed Corportation
SSCA	State Seed Certification Agency
SSR	Simple Sequence repeat
TAMED	N, N, N, N; tetramethylendiamine
TCA	Trichloroacetic acid
TOC	Total organic carbon
TP	Pop of Paper
TP	Turgor pressure
UF	Ultra Filtration
UNDP	United Nations Development Programmee
UPOV	International Union for the Protection of Plant Varieties
UPS & TDC	Uttar Pradesh Seeds and Tarai Development Corporation

Abbreviation	*Full Form*
UWP	Ultimate Wilting Point
VIR	N.I. Vavilow Institute of Plant Industry
VPAS	Vivekanda Parvatiya Krishi Anusandhan Shala
VPKAS	Vivekanand Parvatiya Krishi Anusandhan Shala.
WARDA	West Africa Rice Development Association
WP	Wall Pressure
WTO	World Trade Organization

3

ENGLISH / SCIENTIFIC
NAMES OF PLANTS

English Name	Botanical / Scientific Name
Bamboo fruit	*Bambusa arundinacea*
Bamboo seeds	*Bambusa arundinacea*
Banana	*Musa paradisicum*
Bankulthi	*Atylosia scarabaeoides*
Banti	*Echinochloa stagnina*
Banyan	*Ficus bengalensis*
Barley	*Hordeum vulgare*
Barnyard millet	*Echinochloa frumentacea*
Barnyard	*Echinochloa frumentacea*
Basil	*Ocimum basilicum*
Basket flower	*Centaurea americana*
Bathua	*Chenopodium album*
Bay leaf	*Leurus mobilis*
Beach pea	*Lathyrus maritimus*
Beans, scarlet runner	*Phaseolus coccineus*
Beet (sugar-beet)	*Beta vulgaris*
Beet root	*Beta vulgaris*
Begonia	*Begonia semperflorens*
Belladonna	*Atropa belladonna*
Bells of Ireland	*Molucella laevis*
Bengal gram	*Cicer arietinum*
Ber	*Zizyphus mauritiana*
Bermuda grass	*Cyndoon dactylon*

English Name	Botanical / Scientific Name
Berseem	*Trifolium alexandrinum*
Betel leaves	*Piper betle*
Betel vine	*Piper betle*
Bidroot	*Scirpus grossus*
Bilimbi	*Averrhoa bilimbi*
Bird wood grass	*Cenchrus setigerus*
Bishop's weed or carum	*Trachyspermum ammi*
Bitter gourd	*Momordica charantia*
Black berry	*Rubus fruiticosus*
Black gram	*Phaseolus mungo*
Black-eyed susan	*Thunbergia alata*
Blue lace flower	*Didiscus corerulea*
Blue panic	*Ponicum antidotale*
Blue panicum	*Panicum antidotale*
Blue wishbone flower	*Torenia foumieri*
Bokwa	*Dioscorea pentaphylla*
Borooee	*Gardenia gummifera*
Borooee, Raw	*Gardenia gummifera*
Bottle gourd	*Lagenaria vulgaris*
Bougainvillea	*Bougainvillea glabra, B. spectabilis, B.peruviana, B. buttiana*
Bread fruit	*Artocarpus altilis*
Brinjal	*Solanum melongena*
Broad bean	*Vicia faba*
Broccoli	*Brassica botrytis var. italica*
Browallia	*Browallia viscose*
Brown hemp (Deccan hemp)	*Hibiscus cannabinus*
Brussels sprouts	*Brassica oleracea*
Buck wheat	*Fagopyrum esculentum*
Budhia	*Malothria heterophylla*

English Name	Botanical / Scientific Name
Buffel grass	*Cenchrus ciliaris*
Bullock's heart	*Annona reticulata*
Bulrush millet	*Pennisetum typhoides*
Bur clover	*Medicago hispida*
Butter flower	*Schizanthus* species
Butterfly pea	*Clitoria ternatea*
Cabbage	*Brassica oleracea*
Cacao	*Theobroma cacao*
Calceolaria	*Calceolaria herbeahybrid*
Calendula	*Calendula officinalis*
California poppy	*Eschsholtzia California*
Calliopsis	*Coreopsis bicolor, C. drummondi*
Cambodge	*Garcinia cambogia*
Canary grass	*Phaloris canariensis*
Candy tuft, hardy	*I. sempervirens*
Candy tutt	*Iberis amara, I. Umbellate*
Canna	*Canna indica*
Canna edible	*Canna edulis*
Cape marigold	*Dimorphotheca aurantiaca*
Cape plumbago	*Plumbago capensis*
Caper	*Capparis spinosa*
Carambola	*Averrhoa carambola*
Caraway	*Carum carvi*
Cardamom (large)	*Amomum subulatum*
Cardamom (small)	*Elettaria cardamomum*
Cardinal larkspur	*Delphinium cardinale*
Carnation	*Dianthus caryophyllus*
Carpathian bell flower	*Campanula carpatica*
Carpet grass	*Axonopus affinis*
Carrot	*Daucus carota*

English Name	Botanical / Scientific Name
Cashew fruit	*Anacardium occidentale*
Cashewnut	*Anacardium occidentale*
Cassia	*Cinnamomum aromaticum*
Castor	*Ricinus communis*
Castor bean	*Ricinus communis*
Cathedral bells	*Cobaea scabdens*
Caucasian leopardbane	*Doronicun caucasicun*
Cauliflower	*Brassica oleracea*
Celery	*Apium graveolens*
Cerastium	*Cerastium tomentosum*
Ceylon pasali	*Talinum triangulare*
Chakravarthi Keerai	*Amaranthus* spp.
Chaltha	*Dillania indica*
Chekkur manis	*Sauropus androgynans*
Cherimoyer	*Annona cherimolia*
Cherries red	*Prunus cerasus*
Cherry	*Prunus avium*
Chickling pea	*Lathyrus sativus*
Chicory	*Cichorium intybus*
Chilgoza	*Pinus gerardiana*
Chillies	*Capsicum annum*
Chimti sag	*Polygonum plebijum*
China aster	*Callistephus chinensis*
China pinks	*Dianthus chinensis*
Chinese cabbage	*Brassica pekinesis*
Chinese Forget-me-not	*Cynoglossum amabile*
Cho-cho-marrow	*Sechium edule*
Christmas cherry	*Solanum predocapsicum*
Chrysanthemum	*Chrysanthemum indicum*
Chrysanthemum	*Chrysanthemum* species

English Name	Botanical / Scientific Name
Chumbia	*Dioscorea hamiltonii*
Churkia	*Dioscorea glabra*
Cinchona	*Cinchona ledgeriana*
Cineraria	*Senecio cruentus*
Cinnamon	*Cinnamomum verum*
Citron	*Citrus medica*
Citronella grass	*Cymbopogon winterianus*
Clarkia	*Clarkia elegans*
Clockvine	*Thunbergia alata*
Clove pink	*Dianthus plumarius*
Cloves	*Syzygium aromaticum*
Cluster beans	*Cyamopsis tetragonoloba*
Cocks comb	*Celosis childsi, C. cristata*
Cocao	*Theobroma cocoa*
Coconut	*Cocos nucifera*
Coffee	*Coffea arabica, Coffea canephora*
Coleus	*Coleus blumel*
Colocasia	*Colocasia anti-quorum*
Colocasia	*Colocasia esculenta*
Colocasia stem	*Colocasia antiquorum*
Columbine	*Aquilegia* species
Common millet	*Panicum miliaceum*
Common stocks	*Matthiola incana*
Common vetch	*Vicia sativa*
Coral bells	*Henchera sanguinea*
Cordyline	*Dracaena indivisa*
Coreopsis	*Coreposis lanceolata*
Coriander	*Coriandrum sativum*
Corn flower	*Centaurea cyanus*
Cosmos	*Cosmos bipinnatus*

English Name	**Botanical / Scientific Name**
Cosmos-klondyke	*Cosmos sulphureus*
Cotton	*Gossypium hirsutum, G. barbadense, G. arboreum, G. herbaceum*
Cowage seed	*Mucuna capitata*
Cowpea	*Vigna catjana, V. unguiculata*
Crambe	*Crambe abyssinica*
Crossandra	*Crossandra infundibuliformis*
Cucumber	*Cucumis sativus*
Cumin seeds	*Cuminum cyminum*
Cup and saucer	*Campanula cakycanthema*
Cup and saucer vine	*Cobaea scandens*
Cup flower	*Nierembergia* species
Cuphea	*Cuphea llavea miniata*
Curry leaves	*Murraya koenigii*
Custard apple	*Annona squamosa*
Cyclamen	*Cyclamen africanum*
Cynoglossum	*Cynoglossum amabile*
Cypress vine	*Ipomoea quamoclit*
Dadap	*Erythrina lithosperma*
Daincha seeds	*Sesbania cannabina*
Dallis grass	*Paspalum dialatatum*
Dames Rocket	*Hesperis matronalis*
Date palm	*Phoenix dactylifera*
Dates	*Phoenix dactylifra*
Dattura	*Datura arborea*
Davana	*Artemisia pallens*
Delphinium	*Delphinium elatum*
Dhaincha	*Sesbania* spp.
Dhalia Daisies	*Dhalia* species
Dharaf grass	*Andropogon montanus*

English Name	Botanical / Scientific Name
Dhaulu	*Chrysopogon fulvus*
Dhauns	*Ranatigrina* sp.
Dhurian	*Durio zilethinus*
Dill or Sowa	*Anethum graveolens*
Dinanath grass	*Pennisetum pedicellatum*
Dioscora	*Dioscorea composita, D. floribunda*
Doob	*Cynodon dactylon*
Double beans	*Faba vulgaris*
Dracaena	*Dracaena indivisa*
Drumstick	*Moringa oleifera*
Dusty miller	*Centaurea candidissima*
English daisy	*Bellis perennis*
Epedong sanga	*Peuce danum nagapurense*
Evening scented stocks	*Matthiola bicemis*
Everlasting	*Helichrysum bracteatum*
Falx	*Linum usitattissimum*
Fennel	*Foeniculum vulgare*
Fenugreek	*Trigonella foenum graecum*
Fenugreek seeds	*Trigonella foenum graecum*
Fern asparagus	*Asparagus plumosus*
Fetid cassia	*Cassia tora*
Fever few	*Matricaria* species
Field bean	*Dolichos lablab*
Figs	*Ficus cunia, Ficus carica*
Finger millet	*Eleusine coracana*
Flax-flowering	*Linum grandiflorum*
Four-o-clock	*Mirablis jalapa*
Fox glove	*Digitalis* species
Fox tail millet	*Setaria italica*
Foxglove	*Digitalis lanata*

English Name	Botanical / Scientific Name
French beans	*Phaseolus vulgaris*
Froget-me-not	*Anchusa myosotidiflora*
Gab	*Diospyros embryopteris*
Gaillardia	*Gaillardia* species
Garden cress	*Lepidium sativum*
Garlic	*Allium sativum*
Garmer	*Coleus barbatus*
Geranum	*Geranium* species
Gerusalem cross	*Lychins chalcedonica*
Geum	*Geum* species
Ghosala	*Luffa cylindrical*
Gilia	*Gilia* species
Ginea grass	*Lnicum maximum*
Gingelly seeds	*Sesamum indicum*
Ginger	*Zinziber officinale*
Giria sag	*Suaeda nudiflora*
Gladiolus	*Gladiolus primulinus*
Gladiolus	*Gladiolus* species
Globe amaranth	*Gomphrena globose*
Gloriosa daisy	*Echinacea purpurea*
Gloxima	*Sinnigia speciosa*
Goa beans	*Psophocarpus tetragonolobus*
Godetia	*Godetia grandifkora*
Gogu	*Hisbiscus cannabinus*
Gold dust	*Alyssum saxatile*
Golden cup	*Hunnemannia fumariaefolia*
Goosefoot	*Chenopodium album*
Grape	*Vitis vinifera*
Grape fruit	*Citrus paradisi*
Grass pinks	*Dianthus plumarius*

English Name	Botanical / Scientific Name
Great millet	*Andropogon sorghum*
Greater galangal	*Alpinia galanga*
Green gram	*Phaseolus aureus* Roxb.
Groundnut	*Arachis hypogaea*
Guar	*Cyamopsis tetragonoloba*
Guava	*Psidium guajava*
Guava hill	*Psidium cattleyanum*
Guinea grass	*Pannisetum purpureum*
Gulcharni	*Calonyction muricatum*
Harfarowrie	*Phyllanthus distichus*
Hebiscus	*Hibiscus* species
Heliopsis	*Heliopsis* species
Heliotrope	*Heliopropium* species
Hemp	*Cannabis sativa*
Henna or Mehndi	*Lawsonia inermis*
Hollyhock	*Althaea rosea*
Honesty	*Lunaria annua*
Horse gram	*Dolichos biflorus*
Horse-radish	*Armoracia rusticana*
Hyacinth bean	*Dolichos lablab*
Hyssop	*Hyssopus officinalis*
Impatiens	*Impatiens holsti*
Indian bean	*Lab lab purpureus*
Indian colza	*Brassico campestris*
Indian privet	*Lawsonia alba, L. inermis*
Indian rapeseed	*Brassica campestrie*
Indian squash	*Citrullus vulgaris var. fistulosus*
Indigo	*Indigofera hirsuta*
Ipecac	*Cephaelis ipecacuanha*
Ipomoea	*Ipomoea reptans*

English Name	Botanical / Scientific Name
Isabgol	*Plantago ovata*
Itakian bugloss	*Anchusa italica*
Italian millet	*Setaria italica*
Jack fruit	*Artocorpus heterophyllus*
Jam, safed	*Eugenia malaccensis, Malay apple*
Jambu fruit	*Syzygium cumini*
Japanese Iris	*Iris kaempferi*
Japanese mint	*Mentha arvensis*
Jasmine	*Jasminum officinale, J. grandiflorum, J. sambac*
Jasmine	*Jasminum auriculatum*
Jipoo sanga	*Habenaria cammelinifolia*
Jojoba	*Simmondsia chinensis*
Job's tear millet	*Coix lachryma-jobi*
Job's tears	*Coix lachryma*
Jobs tear	*Coix lacryma-jobi*
Jowar	*Sorgum vulgare/Sorgum bicolour*
Jujube	*Ziziphus mauritiana*
Jungli badam	*Sterculia foetida*
Juniper berry	*Juniperus communis*
Jurmata	*Canthium idymium*
Jute	*Corchorus capsularis*
Kagzi lime	*Citrus aurantifolia*
Kandanthippilli	*Piper longum*
Kankoda	*Momordica dioica*
Kanthankathiri	*Solanum xanthocarpum*
Karonda	*Carissa carandas*
Kathasag	*Dentella repens*
Kazungla	*Setaria sphacelata*
Kenaf	*Hibiscus cannabinus*

English Name	Botanical / Scientific Name
Kenasag	*Commelina benaghalensis*
Kesaur	*Pachyrrhizus angulatus*
Kewara	*Pandanus odoratissimus*
Khamealu	*Dioscorea alata*
Khejri	*Prosopis cineraria*
Kheksa	*Momordica cochinchinensis*
Khesari dhal	*Lathyrus sativus*
Kidney bean	*Phaseolus aconitifolius*
Kila pazham	*Vaccinium leschenaulti*
Kittul flour	*Caryota urens*
Kiwi fruit	*Actinidia deliciosa*
Knol-khol	*Brassica oleracea var. caulorapa*
Kodo millet	*Paspalum scrobiculatum*
Kohar sag	*Bauhinia purpurea*
Koila karhasag	*Astercantha longifolia*
Kokam	*Garcinia indica*
Koo babul	*Leucaena latisiqua*
Korlaleaves	*Bauhinia malabarica*
Korukkapalli	*Pithacellobium dulce*
Kovai	*Coccinia cordifolia*
Koyakeerai	*Amaranthus* spp.
Kuppakeerai	*Amaranthus vindis*
Kuppameni	*Acalypha indica*
Kusum fruits	*Schleichera trijunga*
Ladies finger	*Abelmoschus esculentus*
Lakuch	*Artocarpus lakoocha*
Langsat	*Lansium domesticum*
Lantana	*Lantana camara*
Large canary grass	*Phalaris tuberosa*
Lark spur	*Delphinium ajeicis*

English Name	Botanical / Scientific Name
Lathyrus bean (Khesari dhal)	*Lathyrus sativus*
Leeks	*Allium porrum*
Lemon	*Citrus limon*
Lemon grass	*Cymbopogan flexuosus*
Lemon gum	*Eucalyptus citriodora*
Lemon mint	*Mentha citrata*
Lemon sweet	*Citurs limetta*
Lentil	*Lens esculenta*
Leopards bane	*Doronicum caucasicum*
Lettuce	*Lactuca sativa*
Lettuce tree leaves	*Pisonia alba*
Lichi	*Nephelium litchi*
Lichi, bastard	*Nephelium longana*
Litchi	*Litchi chinensis*
Lima bean	*Phaseolus limensis*
Lime	*Citrus medica var. acida*
Lime	*Citrus aurantifolia*
Lime, sweet Musambi	*Citrus sinensis*
Linaloe	*Bursera delepechianum*
Linaria	*Linaria* species
Linseed	*Linum usitatissimum*
Liquorice or mulhati	*Glycyrrhiza glabra*
Little gourd	*Coccinia indica*
Little millet	*Panicum miliare or Panicum sumatrense*
Lobelia	*Lobelia erinus*
Lokooch	*Artocarpus lakoocha*
Longmelon	*Cucumis melo var. utilissima*
Loquat	*Eriobotrya japonica*
Lotus	*Nelumbium nelumbo or Lotus corniculatum*
Lotus root	*Nelumbium nelumbo*

English Name	Botanical / Scientific Name
Lotus seeds	*Nelumbium nelumbo*
Lovage	*Levisticum officinale*
Love-Lies-Bleeding	*Amaranthus caudatus*
Lucerne	*Medicago sativa*
Lupine	*Lupinus* species
Lupins	*Lupinus* species
Macadamia nut	*Macadamia ternifolia*
Mace	*Myristica fragrans*
Mahua	*Bassia longifolia*
Maiden pinks	*Dianthus deltoids*
Maize	*Zea mays*
Makhana	*Euvvale ferox*
Malmandi	*Idigofera glandulosa*
Maltese cross	*Lychnis chalcedonica*
Manathakkali	*Solanum nigrum*
Mandarin	*Citrus reticulata*
Mango ginger	*Curcuma amada*
Mango green	*Mangifera indica*
Mango steen	*Garcinia mangostana*
Mangosteen	*Garcinia mangostana*
Marguerite, Hardy	*Anthemis kelwayi*
Marigold	*Tagetes erecta, T. patula*
Marigold	*Tagetes* species
Marjoram	*Marjorana hortensis*
Marjorum	*Origanum vulgare*
Marking nut	*Semecarpus anacardium*
Marvel grass	*Dichanthium annulatum*
Marvel of peru	*Mirablis jalapa*
Matasag	*Antidesma diandrum*
Matasura	*Antidesma ghesaembilla*

English Name	Botanical / Scientific Name
Matricaria	*Matricaria inodora*
Mayalu	*Basella rubra*
Melon musk	*Cucumis melo*
Melon water	*Citrullus vulgaris*
Mesta	*Hibiscus sabdariffa*
Methi	*Trigonella foenumgraecum*
Mexican tulip poppy	*Hunnemannia fumariaefolia*
Mignonette	*Reseda odorata*
Mint	*Mentha spicata*
Modakathan keerai	*Cardiospermum helicacabum*
Monarch daisy	*Venidium fastuosum*
Moon flower	*Ipomoea noctiflora*
Moor sanga	*Butea frondosa*
Morning glories	*Ipomoea* species
Moss rose	*Portulaca grandiflora*
Moth bean	*Phaseolus aconitifolius*
Mukarrate keerai	*Boerhaavia repens*
Mulberry	*Morus* sp.
Mulberry	*Morus alba, M. indica, M. serrata, M. laevigata*
Mulchari	*Minusops elengi*
Mullein's pink	*Lychnis coronaria*
Murum sanga	*Dioscorea spinosa*
Muskmelon	*Cucumis melo*
Mustard leaves	*Brassica campestris*
Mustard seed	*Brassica nigra or Brassica juncea*
Napier grass	*Pennisetum purpureum*
Nasturtium	*Trapaeolum* species
Neem fruit	*Malia azadirachta*
Neem tree	*Azardirachta indica*

English Name	Botanical / Scientific Name
Nemesia	Nemesia species
Nemophila	Nemophilia insignis
Nemophila spotted	Nemophilia maculata
Nerringi	Tribulus terrestris
Nicotiana	Nicotiana affinins, N. sanderae
Nierembergia	Nierembergia species
Nigella	Nigella damescena
Niger	Guizotia abyssinica
Niger seeds	Guizotia abyssinica
Nisorha	Cordia dichotoma
Nisorha flowers	Cardia dichotoma
Nutmeg	Myristica fragrans
Oat	Avena sativa
Oatmeal	Avenaby zantina
Ochen sanga	Momordica cochin chinensis
Oil palm	Ealias guinensis
Okra	Abelmoschus escukentus
Omum	Trachyspermum ammi
Onion	Allium cepa
Opium poppy	Papaver somniferum
Orange	Citrus aurantium
Orchard grass	Dactylis glomerata
Oysternut	Telfairea pedata
Pacharisi keerai	Euphorbia hirta
Painted daisy	Pyrethrum species
Palm	Elaeis guineensis
Palmarosa oil grass	Cymbopogon martini
Palmyra fruit	Borassus flabellifer
Panivaragu	Panicum miliaceum
Paniyala	Flacocurtia cataphracta

English Name	Botanical / Scientific Name
Pansy	*Viola tricolor*
Papa	*Gardenia latifolia*
Papaya	*Carica papaya*
Paprika	*Capsicum annuum*
Para grass	*Brachiaria mutica*
Parsley	*Petroselinum crispum*
Parsnip	*Pastinaca sativa*
Paruppu keerai	*Portulaca oleracea*
Parwar	*Trichosanthes dioica*
Parwar sag	*Trichosanther dioica*
Passion fruit	*Passiflora edulis*
Patchouli	*Pogostemon cablin*
Patua sag	*Corchorus capsularis*
Pea	*Pisum sativum*
Pecanut	*Carya illieonsis*
Peach bell flower	*Campanula persicifolia*
Peaches	*Amygdalis persica, Prunus persica*
Pearl	*Achillea ptarmica*
Pearl millet	*Pennisetum americanum*
Pears	*Prunus persica, Pyrus communis*
Persimmon	*Diospyras kaki*
Penstemon	*Pentstemon* spp.
Pepper	*Piper nigrum or Piper longum*
Pepper mint	*Mentha piperita*
Perandai	*Vitis quadrangularis*
Perar	*Randia uliginosa*
Periwinkle	*Vinca rosea*
Persimmon	*Diospyros kaki*
Petunia	*Petunia hybrida*
Phacelia	*Phacelia* species

English Name	Botanical / Scientific Name
Phalsa	*Grewia asiatica*
Phlox	*Phlox drummondi*
Phutka chattoo (Rugroo)	*Lycoperdon* sp.
Physalis	*Physalis* species
Pied gazania	*Gazanis splendens*
Pin cushion flower	*Scabiosa atropurpurea*
Pine apple	*Ananas comosus*
Pink beans	*Phaseolus* sp.
Pipal	*Ficus religiosa*
Pistachio nut	*Pistacia vera*
Piyal	*Buchanania latifolia*
Piyal seeds	*Buchanania latifolia*
Plantain	*Musa sapientum*
Plum	*Prunus domestica*
Plumbago	*Plumbago capensis*
Pomegranate	*Punica granatum*
Ponnanganni	*Alternanthera sessilis*
Poppy	*Papaver nudicaule, P. orientale, P. glaucum, P. rhoeas Papaver somniferum*
Portulaca	*Portulaca grandiflora*
Potato	*Solanum tuberosum*
Primula primrose	*Primula* species
Proso millet	*Panicum miliaceum*
Prunes	*Prunus salicina*
Pummelo	*Citrus maxima*
Pumpkin	*Cucurbita maxima*
Punnaku keerai	*Corchorus acutangulus*
Purple cone flower	*Echinacea purpurea*
Purslane	*Portulaca oleracea*
Pyrethrum	*Pyrethrum* species

English Name	Botanical / Scientific Name
Queen-Anne's -Lace	*Chaerophyllum dasycarpum*
Quince	*Cydonia oblonga*
Radish	*Raphanus sativus*
Ragi	*Eleusine coracana*
Rai	*Brassica juncea*
Raisins	*Vitis vinifera*
Rajagira	*Amaranthus paniculatus*
Rajmah	*Phaseolus vulgaris*
Ramie	*Boehmeria nivea*
Ranunculus	*Ranunculus* species
Rape plant stem	*Brassiea napus*
Rapeseed	*Brassica campestris*
Raspberry	*Rubus wallichi*
Rauwolfia	*Rauwolfia serpentina*
Rayan	*Mimusops hexandra*
Red ants	*Aecophylla smaragdina* fab.
Red gram (pigeonpea)	*Cajanus cajan*
Red hot poker	*Kniphofia* species
Regal lily	*Lilium regale*
Rhodes grass	*Chloris gayana*
Rhubarb stalks	*Rheum emodi*
Rice	*Oryza sativa, Oryza glaberrima*
Ridge gourd	*Luffa acutangula*
Rock cress	*Arabis alpina*
Rocket salad	*Eruca sativa*
Rose	*Rosa multiflora*
Rose apple	*Syzygium jambos*
Rose campion	*Lychnis coronaria*
Rose geranium	*Pelargonium graveolens*
Rosemary	*Rosmarinus officinalis*

English Name	Botanical / Scientific Name
Royal centaurea	*Centaurea imperialis*
Rozelle	*Hibiscus sabdariffu*
Rubber	*Hevea brasiliensis*
Rye	*Secale cereale*
Rye grass	*Lolium perenne*
Safflower	*Carthamus tinctorius*
Saffron	*Crocus sativus*
Sage	*Salvia officinalis*
Sain	*Sehima nervosum*
Salpigossis	*Salpiglossis gloxinaeflora, S. sinuate*
Samai	*Panicum miliare*
Sanga-ka-phal	*Dioscorea puber*
Sannhemp	*Crotalaria juncea*
Sanvitalia	*Sanvitalia procumbens*
Sanwa millet	*Echinochloa frumantacea*
Saponaria	*Saponaria ocymoides, S. vaccaria*
Sapota	*Achras sapota*
Saravallai keerai	*Trianthema monogyna*
Sarli sag	*Vangueria spinosa*
Savory	*Satureja hortensis*
Scabiosa perennial	*Scabiosa caucasia*
Scarlet sage	*Salvia splendens*
Schizanthus	*Schizanthus* species
Seemai ponnanganni	*Alternanthere* sp.
Seethaphal	*Annona squamosa*
Senji	*Melilotus parviflora*
Senna	*Cassia angustifolia*
Sensitive plant	*Mimosa pudica*
Sesamum	*Sesamum indicum*
Setaria grass	*Setaria anceps*

English Name	*Botanical / Scientific Name*
Shaftal	*Trifolium resupinatum*
Shasta daisy	*Chrysanthemum maximum, C. lencanthemum*
Shepu	*Peucedanum graveolens*
Shevri	*Sesbania aegyptiaca*
Siberian wall flower	*Cheiranthus cheiri, C. allioni*
Silk cotton flowers	*Bombax malabaricum*
Sinduar sag	*Celosia argentia*
Sinduar sag wild	*Allmania polygonoides*
Siratro	*Macroptilum atropurpureum*
Siris	*Albizia lebbeck*
Sirka	*Zizyphus rugosa*
Sirukeerai	*Amaranthus polygonoides*
Sisal	*Agave sisalana*
Snake gourd	*Trichosanthes anguina*
Snapdragon	*Antirrinum* species
Snap-melon	*Cucumis melo*
Sneezeweed	*Helenium* species
Snow in summer	*Cerastium tomentosum*
Snow on the Mountain	*Euphorbia marginata*
Solanum	*Solanum* species
Sonchal sag	*Malva parviflora*
Song	*Dioscorea anguiera*
Sorghum	*Sorghum bicolour*
Soyabean	*Glycine max*
Spearmint	*Mentha spicata*
Spider plant	*Celome gigantean*
Spiked millet	*Stapf and Hubbard*
Spinach	*Spinacia oleracea*
Sponge gourd	*Luffa cylindrica*

English Name	Botanical / Scientific Name
Sprenger asparagus	*Asparargus sprengeri*
Star anise	*Illicium verum*
Star apple	*Eugenia javanica*
Statice	*Statice sinuate*
Stock	*Mathiola incana*
Straw flower	*Helichrysum monstrosum*
Strawberry	*Fragaria vesca*
Stylos	*Stylosanthes humilis*
Sudan grass	*Sorghum sudanense*
Sugarcane	*Saccharum officinarum*
Summer cypress	*Kochia childsii*
Summer squash	*Cucurbita pepo*
Sun flower	*Helianthus* species
Sundakai	*Solanum torvum*
Sunflower	*Helianthus annus*
Sunn hemp	*Crotolaria juncea*
Susni sag	*Marsilea minuta*
Sutari	*Phaseolus calcaratus*
Swan river daisy	*Brachycome iberiddifolia*
Sweet cherry	*Prunus avium*
Sweet flag or bach	*Acorus calamus*
Sweet orange	*Citrus sinensis*
Sweet pea	*Lathyrus odotatus*
Sweet potato	*Ipomoea batatas*
Sweet Rocket	*Hesperis matronalis*
Sweet sultan	*Centaurea moschata*
Sweet william	*Dianthus barbatus*
Sweet-wivels field	*Dianthus chinensis*
Sword beans	*Canavalia gladiata*
Tahoka daisy	*Machaeranthera tanacetifolia*
Tall fescue	*Festuca aurandinacea*
Tamarind	*Tamarindus indica*
Tapioca	*Manihot esculenta*

English Name	Botanical / Scientific Name
Taramira	*Eruca sativa*
Tarragon	*Artemisia dracunculus*
Tea	*Camellia sinensis*
Tea	*Camellia thea, C. chah*
Tejpat	*Cinnamomum tamala*
Teosinte	*Euchlaena mexicana*
Tetrolobar bean	*Lotus tetragonolobus*
Texas blue bonnet	*Lupinus subcamosus*
Thavittupazham	*Rhodomyrtus tomentosa*
Thrift	*America* species
Thunbergia	*Thunbergia alata*
Thyme	*Thymus vulgaris*
Tidy tips daisy	*Layia elegans*
Tinda	*Citrullus vulgaris*
Tithonia, Torch flower	*Tithonia speciosa*
Tobacco	*Nicotiana tabacum, N. rustica*
Tobacco flowering	*Nicotiana affinis*
Tomatillo	*Physalis ixocarpa*
Tomato	*Lycopersicon esculentum*
Torchilly, Tritoma	*Kniphofia* species
Torpedo grass	*Panicum repens*
Transvaal daisy	*Gerbera jamesoni*
Tree tomato	*Cyphomandra betacea*
Tuberose	*Polianthes tuberosa*
Tuki	*Diospyros melanoxylon*
Turmeric	*Curcuma domestica, C. longa*
Turnip	*Brassica rapa*
Turum sanga	*Curculigo orchioides*
Vanilla	*Vanilla planifolia*
Varagu	*Paspalum scrobiculatum*
Veethi keerai	*Cadalia indica*
Vegetable marrow	*Cucurbita pepo*
Velai keerai	*Hydrolea* sp.

English Name	Botanical / Scientific Name
Vella keerai	*Cleome viscose*
Velvet bean	*Mucuna cochinchinensis*
Velvet centaurea	*Centaurea gymnocarpa*
Verbena	*Verbena hybrida*
Vetches	*Vicia* species
Vetiver or Khus	*Vetiveria zizanioides*
Vikki pazham	*Elaeo carpus oblongus*
Vilaiti babool	*Acacia tortilis*
Vinca	*Vinca rosea*
Viola	*Viola cornuta*
Virginian stocks	*Malcolmia maritime*
Wallflower	*Cheiranthus allioni, C. cheiri*
Walnut	*Jugans regia*
Water chestnut	*Trapa bispinosa*
Water cress	*Nasturtium oficinale*
Water lily	*Nymphea nouchali*
Watermelon	*Citrullus vulgaris*
Weeping love grass	*Eragrostis curvula*
Wheat	*Triticum aestivum, T. durum, T. dicoccum, T. sphaerococcum*
Willow leaf ox eye	*Bupthalmum salicifolium*
Winter squash	*Cucurbita maxima*
Wood apple	*Limonia acidissima or Feronia limonica*
Wood sand piper	*Tringa galareola*
Yam	*Dioscorea alata*
Yam ordinary	*Typhonium trilobatum*
Yam wild	*Dioscorea versicolor*
Yam, elephant	*Amorphophallus campanulatus*
Yeast	*Terula saccharomyces, S. cerevacies*
Yucca	*Yucca filamntosa*
Zinnina	*Sanvitalia procumbens, Zinnia elegans, Z. grandiflore, Z. haageana, Z. linearis*
Zizyphus	*Zizyphus jujuba*

4

ENGLISH / SCIENTIFIC NAMES OF FISH, POULTRY AND OTHERS

English Name	Scientific Name
Air	*Mystus seenghala*
Anchovy	*Engraulis mystax*
Bacha	*Eutropiichthys vacha*
Bam	*Mastocembellus armatus*
Baspata machli	*Ailia coilia*
Bata	*Chondrostoma gangeticum*
Beef	*Bos taurus*
Bele	*Glassogobius giuris*
Bhangan bata	*Labeo bata*
Bhanger	*Mugil tade*
Bhekti	*Lates calcarifer*
Bhole	*Serranus lanceolatus*
Big-jawed jumper	*Lactarius lactarius*
Black rat	*Rattus rattus*
Blue mussel	*Mytilus viridis*
Boal	*Wallago attu*
Bombay duck	*Harpadon nehereus*
Brown rat or Norway rat	*Rattus poryegicus*
Buffalo	*Bulbus bubalis*
Cat fish	*Arius sona*
Chela	*Chkela phulo*
Chingri goda	*Macrobrachium rudis*
Chingru	*Paloemus carcinus*

English Name	Scientific Name
Chital	*Notopterus chitala*
Confused flour beetle	*Tribolium confusum*
Crab	*Paratephusa spinigera*
Duck	*Anasppla tyrhyncha*
Fig moth	*Ephestia cautella*
Finch	*Fring illidoe*
Folui	*Notopterus notopterus*
Fowl	*Gallus bankiva nurghi*
Grain moth	*Sitotraga cerealella*
Ghol	*Sciane miles*
Goat	*Capra hyrchusb*
Goggler	*Caranx crumenophthalmus*
Granary weevil	*Sitophilis granarius*
Grey quail	*Cotuenic coturnix*
Herring, ox-eyed	*Megalops cyprinoids*
Hilsa	*Clupea ilisha*
Horse mackerel	*Caranx melampygus*
House mouse	*Mus musculus*
Indian whiting	*Sillago sihama*
Jew fish (Kora)	*Pseudosciaena coibor*
Jew fish (Pallikora)	*Octolithes rubber*
Kalabasu	*Labeo calbasu*
Katla	*Catla catla*
Khapra beetle	*Trogoderma granarium*
Khorsula	*Mugil corsula*
Khoyra	*Gonialosa manminna*
Koi	*Anabas testudineus*
Koocha machli	*Amphipnous cuchia*
Larger bandicoot	*Bandicoot indica*
Lata	*Ophioce phallus punctatus*

English Name	Scientific Name
Lesser bandicoot	*Bandicoot Bengal ensis*
Lesser grain borer	*Rhizopertha dominica*
Lobster	*Palae mon* sp.
Mackerel	*Rastrelliger kanagurta*
Magur	*Clarias batrachus*
Mahasole	*Barbus tor*
Mandeli	*Coilia dussumieri*
Mrigal	*Cirrhinus mrigala*
Mullet	*Mugil oeur*
Oil sardine	*Sardinella longiceps*
Pabda	*Callichorus pabo*
Pangas	*Pangasius pangasius*
Parsey	*Mugil parsia*
Pigeon	*Columba livia intermedia*
Pomfret black	*Formio niger*
Pomfret white	*Stromateus sinensis*
Pork	*Sus cristatus wagner*
Prawn	*Penaeus* sp.
Pulse beetle	*Collosobruchus* spp.
Puti	*Burbus* sp.
Ravas	*Polynemus tetradactylus*
Ray	*Rhinoptera sewelli*
Red flour beetle	*Tribolium cost anem*
Ribbon fish	*Trichiurus* sp.
Rice moth	*Corcyra cerealella*
Rice weevil	*Sitophilus oryzae*
Rohu	*Labeo rohita*
Ruff and Reeve	*Philomachus pugnax linn*
Sardine	*Sardinella fimbriata*
Sarputi	*Barbus sarana*

English Name	Scientific Name
Saw toothed grain beetle	*Oryzaephilus surinamensis*
Seer	*Cybium guttatum*
Shark	*Carcharias* sp.
Silver belly	*Leiognathus insidiator*
Singhala	*Arius dussumieri*
Singhi	*Saccobranchus fossilis*
Snail, big	*Pila globosa*
Snail, small	*Viviparus bengaensis, F. typical*
Sole	*Pohiocephalus striatus*
Sole (malabar)	*Cynoglossus semifaciatus*
Surmai	*Cybium commersoni*
Tapsi	*Polynemus paradiseus*
Tartoor	*Opisthopterus tardoore*
Tengra	*Mystus vittatus*
Tunny	*Thynnus macropterus*
Venison	*Antilope cervicapra* Linn.
White bait	*Anchoviella* sp.

5

TERMINOLOGY

Term	Definition
α-Amylase	An enzyme with endo – (1, 4) – glucanase activity. This enzyme is used commericially in food preparations, thining of starch for alcohol fermentations.
β-Oxidation pathway	Fatty acid breakdown involves a repeating sequence of four reactions.Oxidation of the acyl CoA by FAD to form a trans-D^2-enoyl CoA; Hydration to form 3-hydroxyacyl CoA; Oxidation by NAD^+ to form 3-ketoacyl CoA; Thiolysis by a second CoA molecule to form acetyl CoA and an acyl CoA shortened by two carbon atoms. The $FADH_2$ and NADH produced feed directly into oxidative phosphorylation, while the acetyl CoA feeds into the citric acid cycle where further $FADH_2$ and NADH are produced. In animals the acetyl CoA produced in β–oxidation cannot be converted into pyruvate or oxaloacetate and cannot therefore be used to make glucose. However, in plants two additional enzymes allow acetyl CoA to be converted into oxaloacetate via the glyoxylate pathway.
'A' line	The male sterile parent in a cross being made to produce hybrid seed (use male sterile line).
'B' line	The fertile counterpart parent of the 'A' line, the maintainer line is known as 'B' line.
2, 4, 5 – T.	2, 4, 5-trichlorophenoxyacetic acids, a translocated hormone weedkiller used to control scrub and woody vegetation.
2, 4-D.	2, 4-dichlorophenoxyacetic acids, a synthetic translocated hormone weed killer used to control broad leaved herbaceous plants and as a defoliant.
Abietic acid	A dichlorophenoxyacetic acid a synthetic translocated hormone weed killer used to control broad leaved herbaceous plants and as a defoliant.

Term	Definition
Abiotic	Pertaining to nonliving environmental factors, such as temperature, light, water and nutrients.
Abnormal seedlings	Those seedlings, which do not show the capacity for continued development into normal plants when grown in good quality soil, and under favourable conditions of water supply, temperature and light.
Abscisic acid (ABA)	A plant hormone that generally acts to inhibit growth, promote dormancy, and help the plant withstands stressful conditions.
Abscission	The shedding of leaves or fruits as a result of chemical activity of cells in abscission zone that extends across the base of the petiole premature formation of abscission layers is a hyperplastic symptom of some plants.
Absorbance	A measure of the extent to which light is attenuated during passage through a coloured liquid or solid.
Absorption spectrum	The range of a pigment's ability to absorb various wavelengths of light.
Absorption spectrum	The image recorded when electromagnetic radiation from a source emitting a continuous spectrum is passed through a substance. If the material is in the gaseous phase bands will appear in the same position as the lines that occur in the charactristic emission spectrum of that substance.
Abysal zone	The portion of the ocean floors where light does not penetrate and where temperatures are cold and pressures intense.
Acclimation	Physical adjustment to a change in an environmental factor.
Acetaldehyde	An aldehyde formed from the oxidation of ethanol. An aldehyde contains a carbonyl group having only one hydrogen attached (formaldehyde, $H_2C=O$, is an exception).
Acetone	One of the ketone bodies that is present normally in the blood in small amounts but rises to toxic levels when there is not enough oxaloacetate to keep the TCA cycle functioning.

Term	Definition
Acetyl CoA	The entry compound for the Krebs cycle in cellular respiration, formed from a fragment of pyruvate attached to a coenzyme.
Acetyl-CoA	A two-carbon (acetyl) molecule made active by the attachment of coenzyme A. Acetyl-CoA is in a key position in the reclaiming of energy from fuel nutrients or in the synthesis of lipids derived from the consumption of excess fuel nutrients.
Achene	A small dry, hard, one-chambered one-seeded indehiscent fruit, as in buckwheat, lettuce and spinach, commonly referred to as a seed.
Acid	According to the Bronsted definition, any ion or molecule, which can furnish a proton to solution?
Acid-base balance	An optimal balance between the number of acidic ions and basic ions in the fluid of a tissue. This is a requirement for the continuation of life process since vital reactions can take place only when the pH is maintained within narrow limits. Both the respiratory and renal systems contribute to the maintenance of the acid-base balance by excreting or conserving the acidic or basic ions under their control.
Acidic amino acids	Amino acids whose side chains have an extra carboxyl (–COOH) group.
Acidosis	A set of conditions that would tend to lower the pH of blood if the body did not act to resist this change. Metabolic acidosis refers to changes in the bicarbonate concentration of the plasma; respiratory acidosis refers to changes in the carbon dioxide pressure.
Acquired character	A change from the normal type brought about by environmental influence.
Acquired immunity	The type of immunity achieved when antigens enter the body the body naturally or artificially. Acquired immunity is due to stimulation of antibody production and production of memory cells keyed to the antigen.
Actin	A globular protein that links into chains, two of which twist helically about each other, forming microfilaments in muscle and other contractile elements in cells.

Term	Definition
Activation energy	The energy, Ea, defined by the Arrhenius equation. $R = Ae^{-Et}/RT$, which is interpreted as the difference between the average energy of the reactant molecules and the energy required to form the reactive intermediate (activated complex) which leads to product.
Activator	Some enzymes have binding sites for small molecules that stimulate their activity; these stimulator molecules are called activators.
Active site	A region on the surface of an enzyme which specifically binds the substrate and which facilitates the reaction of the substrate to form product.
Active transport	Active transport of a molecule across a membrane against its concentration gradient requires an input of metabolic energy. In the case of ATP-derive active transport, the energy required for the transport of the molecule (Na^+, K^+, Ca^{++} or H^+) across the membrane is derived from the coupled hydrolysis of ATP (*e.g.*, Na^+/K^+- ATPase). If both the molecule to be transported and the ion move in the same direction across the membrane, the process is called symport (*e.g.*, Na^+/glucose transporter). If the molecule and the ion move in opposite directions it is called antiport (*e.g.*, erythrocyte band anion transporter).
Active transport	Energy requiring transport of a solute across a membrane in the direction of greater concentration or in other words, against a concentration gradient. The usual, unassisted movement of molecules is by diffusion from a place of higher concentration to a place of lower concentration. The work of active transport to maintain concentration gradient in the cells of a human at rest is so great that it may consume as much as 30-40 per cent of the total input of energy.
Active transport	A biological mechanism that results in the transport of molecules across a membrane against a concentration gradient. Active transport requires metabolic energy and may be associated with a specific carrier protein or lipoprotein molecule.

Term	Definition
Active transport (pump)	Protein molecules in the cell membrane transport ions or molecules through membrane; movement may be against concentration gadient (*i.e.,* from region of lower to region of higher concentration of concerned molecules, and is supported by ATPase activity).
Activity	The "thermodynamically effective" concentration of a substrate. Its value depends upon the choice of a standard state for the substrate referred to. For solutes, the standard state is often taken as 1 M concentration, and the activity of the solute is approximately equal to the molar concentration in vary dilute solutions. For solvents the standards state is taken as the pure solvent, and the activity of a solvent in dilute solution is approximately unity.
Activity	The "thermodynamically effective" concentration of a substrate.
Acyl carrier protein	A protein that has acid as its prosthetic group. It is active in the synthesis of fatty acids as it carries acyl groups that are bound to the pentothenic acid in a thioester linkage.
Adaptation	Modification of an organism fitting it more perfectly to conditions of its environment.
Adenine/Guanine	Nitrogen base one of two purines found in both DNA and RNA.
Adenosine triphosphate (ATP)	The main energy currency for cells. ATP energy is used to promote ion pumping enzyme activity and muscular contraction.
Adhesion, cohesion and surface tension	The attraction between two dissimilar molecules is known as adhesion and between two similar molecules as cohesion. Cohesion in a liquid tends to prevent escape from the surface. This results in surface tension. The surface tension varies from liquid to liquid. In any liquid the surface tension is inversely related to the temperature.
Adipsin	A protein that appears to be made adipose cells and acts as a communication link between adipose cells and the brain.
Admixture	Seeds or other matter, other than the kind of seed specified.

Term	Definition
Adsorption	In physical chemistry, the adhesion of molecules to solids. This process forms the basis of adsorption chromatography, which is used to purify proteins and other macromolecules as well as some method of cell or enzyme immobilization.
Aeration	The moving of air through seed at low airflow rates for purposes other than drying.
Aerobe	An organism that requires oxygen for respiration and hence growth.
Aerobic reaction	A chemical reaction requiring oxygen.
Aerobic respiration	Harvesting chemical energy in the form of ATP from food molecules, with oxygen as the final electron acceptor.
AIDS	The name of the late stage of HIV infection; deficiency by a specific reduction of T cells and the appearance of characteristic secondary infections.
Air screen cleaner	It is the basic seed-processing machine. An air screen does basic cleaning of all seeds cleaner.
Albinism	A rare recessive disorder in which the enzyme tyrosinase is missing in the pigmented cells.
Albumin	A serum globular protein that occurs in the highest concentration of the major serum protein.
Alcohol	A hydrocarbon in which a hydroxyl group has replaced one or more of the hydrogen.
Aldehyde	An organic molecule with a carbonyl group located at the end of the carbon skeleton.
Aleurone	Protein matter in the form of grains occurring in the endosperm of ripe seeds.
Aliphatic	Belonging to that series of organic compounds characterized by open chains of carbon atoms rather than by rings.
Alkalosis	A set of conditions that would tend to raise the pH of the blood if the body did not act to resist this change.
Allele	One of two normally alternate forms of a gene one being the normal wild type (+), the other the mutant type.

Term	Definition
Allopolyploid	A common type of polyploid species resulting from two different species interbreeding and combining their chromosomes.
Alpha (α) helix	A spiral shape constituting one form of the secondary structure of proteins, arising from a specific hydrogen-bonding structure.
Amino acid	The building block for proteins containing a central carbon atom with a nitrogen atom and other atoms attached OR an organic molecule possessing both carboxyl and amino groups. Amino acids serve as the monomers of proteins.
Amino acid codons	Val = GUA, Phe = UUU, Lys = AAA, Gly = GGA, Ser = UCC, Arg = AGA, Pro = CCC, Glu = GAA, Leu = UUG, Ala = GCA, Trp = UGG, Met = AUG.
Amino acid pool	The free amino acids present mainly in the cytosol and circulating blood. These amino acids may have joined the pool during either the digestion and absorption of protein rich foods or the degradation of body proteins. The amino acids stand ready to be incorporated into protein or used for fuel.
Ammonia nitrogen	The ammonia nitrogen is a result of bacterial decomposition of organic matter. Fresh sewage is generally high in organic nitrogen and low in ammonia nitrogen. The sum of organic and ammonia nitrogen should remain constant for the same liquid wastes, unless ammonia is allowed to escape to the atmosphere because of septic action. The total concentration of the two serves as a valuable index for evaluating the strength of liquid waste and for determining the type of treatment process to select.
Ammonotelic	Organisms that excrete nitrogen as ammonia.
Anabolic pathway	A metabolic pathway that consumes energy to build complicated molecules from simpler ones.
Anabolic reactions	The reactions by which the structural and functional components of the cell are synthesized.
Anabolism	The energy requiring production of macro-molecules from smaller molecules.
Anaerobic reaction	A chemical process that does not require oxygen.

Term	Definition
Anaerobic respiration	A form of respiration that occurs in a few groups of bacteria living in anaerobic environments such as soil, the final electron acceptors are sulfate and nitrate.
Analbuminemia	A deficiency in the livers capacity to synthesize the protein albumin thus leading to a decrease in the plasma albumin.
Analogy	The similarity of structure between two species that are not closely related; attributable to convergent evolution.
Anaphase	A stage in cell division following metaphase in which the daughter chromosomes, pass from the equatorial plate in metaphase midway to the two poles at the opposite ends of the cell.
Angstrom (Å)	A unit of length equal to 10^{-8} cm.
Anhydride	A compound formed by the loss of water between two acids
Anion	A negatively charged ion.
Anther	In seed plants the part of the stamen where pollen grains (microspores) are formed.
Anthesis	The process of dehiscence of anthers, the period of pollen distribution.
Anthocyanin	A soluble glucoside pigment producing either reddish or purplish colour to flowers and other parts of plants.
Antibiotic	A chemical that kills or inhibits the growth of bacteria, often via transcriptional or translational regulation.
Antibody	An antigen-binding immunoglobulin produced by B cells, that functions as the effectors in an immune response; or the capacity of proteins to distinguish among different molecules or immunoglobulins than in enzymes. Antibodies are valuable tools for identifying and purifying proteins.
Antibody	A serum protein that is synthesized in response to the entry of a foreign substance (antigen).
Anticodon	A specified base triplet on one end of a tRNA molecule that recognizes a particular complementary codon on an mRNA molecule.

Term	Definition
Anticodon	A sequence of three bases in transfer RNA that is complementary to a sequence of three bases in messenger RNA.
Antigen	A substance (usually a protein) capable of stimulating antibody production when introduced into an animal body.
Antimetabolite	A substance that bears a structural resemblance to a normal substrate or enzyme and thus competes with it in metabolism.
Antioxidant	An agent that prevents or inhibits oxidation of a substance by combing with oxygen.
Apoenzyme	The inactive protein part of an enzyme that remains after the cofactor is removed.
Apoenzyme	The enzyme or protein part of an enzyme cofactor complex that has lost the cofactor and is thus inactive.
Apoferritin	A protein in the intestinal cell that binds with the ferric form of iron (Fe^{+3}) to form ferritin.
Apomixis	Peprodution without sexual union, unfertilized egg or from somatic cells associated with the egg.
Apoplast	Part of the plant tissue matrix outside the membranes enclosing the cell contents, *i.e.,* outside the plasmalemma.
Aqueous solution	A solution in which water is the solvent.
Ascopore	A spore produced by free-cell formation following meiosis in an ascus.
Ascostroma	A stromatic ascocarp bearing asci directly in locules within the stroma.
Ascus	A sac like cell of the Ascomycota in which karyogamy is followed immediately by meiosis and in which ascopores of definite number arise by free cell formation
Aseptic	It is a techniques designed to inhibit actively putrefactive microorganisms by extension of techniques designed to inhibit or excelude all microorganisms.

Term	Definition
Asexual	Of structures and reproductive processes lacking gametes for conjugation or fertilization.
Aspartame	L-Aspartyl-Phenylalanine methyl ester; a colourless crystalline material that is 160 times sweeter than sucrose.
Aspirator	A seed processing machine in which seed separations are made by use of air, on the basis of differences in terminal velocity of seed. It has a fan, at the discharge point, which creates a vacuum or negative pressure within the machine. Air rushing to fill in the vacuum creates a stream of air, which is used to separate seeds.
Aspsis	The absence of microorganisms that produce disease or contaimination of other culture or process characterized by the used of feedstocks and other materials that are free from such organisms. Asepsis may be achived by strilization of media and equipment.
Assimilation	A general term for constructive metabolism; usually of carbon assimilation (photosynthesis). The conversion of simple molecules into the complex constituents of the living matter.
Assimilation number	It is the amount of CO_2 absorbed in grams per hour for each gram of chlorphyll during photosynthesis.
Asymmetric carbon	A carbon atom covalently bonded to four different atoms or groups of atoms.
Atheroclerosis	Atherosclerosis is characterized by cholesterol rich arterial thickenings (atheromas) that narrow the arteries and cause blood clots to form. If these blood clots block the coronary arteries supplying the heart, the result is a myocardial infarction or heart attack.
Atmosphere	Gaseous layer enveloped by the earth is called atmosphere.
Atmosphere	Gaseous layer enveloped by the earth is called atmosphere.
Atom	One of the smallest parts of an element that can exist.

Term	Definition
Atomic Absorption Spectroscopy	A method used for the quantitive determination of metals. The sample is burnt in a flame through which ultraviolet light generated from a lamp specific for the element under invetigation is passed. Attenuation of the beam is detected using a photomultiplier and converted into an electrical signal, which is concentration or amount using a microprocessor.
Atomic number	The number of protons in the nucleus of an atom, unique for each element and designated by a subscript to the left of the elemental symbol.
Atomic weight	The total atomic mass, which is the mass in grams of one mole of the atom.
Attenuation	A regulatory mechanism employed by bacterial cell. Whereas enzyme repression allows the cell to respond to extreme concentrations of metabolites, it probably represents a means of fine-tuning to relatively mild fluctuatings in the concentrations of metabolities.
Authority	The name of the author who was the first to give a species or other taxon to its name. In the case of species the authority is given after the binomial.
Auto ecology	It is study of inter-relationship between individual species or its population and its environment.
Autoanalyzer	Registered name applied to fully automated analytical system marketed by Technicon. The autoanalyzer is based on automatic sampling and continuous flow through a variable manifold to a detector (colorimeter, absorptiometer, fluorometer, etc.). Bubbles of gas keep the samples separately.
Autogamy	Self-fertilization found in some flowers and in some gametes derived from the same gametangium.
Autolysis	The breakdown of animal or plant products as result of the action of enzyme contained within the cells or tissue affected.
Autoradiography	A technique for detecting the location of a radioisotope in a tissue, cell or molecule. The sample is placed in contact with a photographic emulsion, usually an X-ray film.

Term	Definition
Autosome	Those chromosomes of the dividing nucleus carrying genes for any factor except that of sex.
Autosomes	Chromosomes other than sex chromosomes.
Autospore	A nonmotile asexual daughter cell, which lacks the ontogenetic potentiality of motility.
Autotrophic nutrition	A mode of obtaining organic food molecules without eating other organisms. Autotrophus use energy from the sun or from the oxidation of inorganic substances to make organic molecules from inorganic ones.
Autotrophic plant	A self-nourishing plant; one capable of making its own food.
Autotrophy	Refers to the ability to synthsize all organic components from simple inorganic compounds like carbon dioxide, ammonia, nitrate and sulfate.
Auxins	A class of plant hormones, including indolacetic acid, that has a variety of effects, such as phototropic response through stimulation of cell elongation, stimulation of secondary growth and development of leaf trances and fruit.
Avidin	A protein found in raw egg whites that can bind biotin and inhibit its absorption. Cooking destroys avidin.
Avidin	A glycoprotein found in raw egg whites that binds biotin in the intestinal tract and inhibits the absorption of biotin.
Avogadro's law	The law states that equal volume of different gases under the same temperature and pressure contains the same number of molecules.
B-cell	A type of lymphocyte that develops in the bone marrow and later produces antibodies, which mediate humoral immunity.
Back cross	The cross of a hybrid to one of the parental types.
Bag closer	The machine used for sewing filled seed bags.
Bagger weigher	These are the small machines which, when properly mounted beneath a bin, will fill and weigh a bag in a single operation.

Term	Definition
Barr body	The dense object that lies along the inside of the nuclear envelope in cell of female mammals, representing the one inactivated X-chromosome.
Basal body	A cell structure identical to a centriole that organizes and anchors the microtubule assembly of a cilium or flagellum.
Basal metabolic rate (BMR)	The rate at which the body uses oxygen when it is at rest and food has not been ingested for 12 hours.
Base	According to the Bronsted definition, any ion or molecule, which can accept a proton.
Basic amino acids	Amino acids whose side chains have positively charged amino groups.
Basic cleaning	Refers to the separations of material larger and smaller than good seed; general size grading and cleaning, such as cleaning by an air screen machine.
Basic seed	A class of seed in a seed certification programmes that is the last step in the intital seed multiplication and is intended for the production of certificed seed. Seed Stock used for the same purpose as basic seed, but not under a certification programme is referred to as the equivalent of basic seed.
Batch	A quantity of harvested crop put into bin or container on a repetitive basis, specifically for treatment such as drying.
Berry	A simple fleshy fruit.
Beta (β) plated sheet	A zigzag shape constituting one form of the secondary structure of proteins formed of hydrogen bonds between polypeptide segments running in opposite directions.
Beta carotene ($C_{40}H_{56}$)	A carotenoid hydrocarbon pigment found widely in nature, always associated with chlorophyll; converted to vitamin A in the liver of many animals.
Beta oxidation	The enzymatic process that occurs in mitochondria whereby Fatty acids are oxidatively degraded at the second carbon atom from the carboxyl group.

Term	Definition
Biennial	The kind of plant that produces vegetable growth during the first year of growing season. After a period of storage or over-wintering out of doors, flowers, fruits and seeds are produced during the second year and the plant dies.
Bile pigment	A breakdown product of the heme portion of hemoglobin, myoglobin and cytochromes.
Bile salts	Bile salts (bile acids) are the major excretory form of cholesterol. These polar compounds are formed in the liver by converting cholesterol into the activated intermediate cholyl CoA and then combining this compound with either glycine to form glycocholate or taurine to form taurocholate. The detergent-like bile salts are secreted into the intestine where they aid the digestion and uptake of dietary lipids.
Bile salts	Derivatives of cholesterol those are synthesized in the liver and stored in the gallbladder.
Bilirubin	A bile pigment resulting from the breakdown of heme, the non-protein iron-containing portion of hemoglobin, myoglobin, and cytochromes.
Biliverdin	A product of the first step in the degradation of the heme portion of hemoglobin, myoglobin and cytochromes.
Bin	An enclosed structure used for storage of seeds.
Bin sampler	The large sized trier, used for drawing samples from bins. These are constructed on the same principles as bag triers but are much larger.
Bioassay	It is the use of an organism (or living system) for assay purposes.
Bioautography	A method for detecting small amounts of one or more substances in a complex mixture, such substances being essential growth requirements for the test organisms used.
Bioavailability	The degree to which the amount of an ingested nutrient actually gets absorbed and so is available to the body.

Term	Definition
Biocatalyst	A catalyst that consits of or is derived from a living organism or tissue or cell culture.
Biochemical genetics	The branch of genetics concerned with the inheritance of genetic differences in the ability or inability to synthesize or metabolize certain chemicals.
Biochemistry	Biochemistry is defined as it deals with chemical processing in living matter from smallest to biggest organisms.
Biological value	The body's ability to retain nutrient absorbed from a food.
Biome	Term used for large regional or subcontinental biosystem characterized by a major vegetation type or other identifying landscape aspect.
Biome	Term used for large regional or sub-continental biosystem characterized by a major vegetation type or other identifying landscape aspect.
Biosynthesis	The synthesis of orgnic molecules by living organisms using linked rections in which the energy and reducing power are derived from ATP and reduced pyridine nucleotides generated by fermentation, respiration, chemosynthesis or photosynthesis.
Biotechnology	The use of advanced scientific techniques to alter and ideally improve characteristics of animals and plants.
Biotic	Pertaining to the living organisms in the environment such as pests, insects, birds, bacteria fungus etc.
Biotin	A vitamin of the B complex that functions as a coenzyme in some reactions associated with the fixation of carbon dioxide.
Biotype	A population of individuals with indentical genetic constitution. A biotype may be homozygous or heterozygous.
Black tongue	A disease seen in dogs that is an animal analog of pellagra, the niacin deficiency disease in humans.

Term	Definition
Blood brain barrier	A specialized capillary arrangement in the brain that restricts the passage of most substances into the brain, thereby preventing dramatic fluctuations in the brain's environment.
Blood pressure	The hydrostatic force that blood exerts against the wall of a vessel.
Bolt	Formation of an elongated stem or seed stalk. In the case of bennial plants, this generally occurs during the second season of growth.
Bond energy	The quantity of energy that must be absorbed to break a particular kind of chemical bond equal to the quantity of energy the bond releases when it forms.
Border rows	The recommended number of rows of the male parental line grown on all the sides of hybrid seed field growning two different parents.
Bottleneck effect	Genetic drift resulting from reduction of a population, typically by a natural disaster, such that the surviving population is no longer genetically representative of the original population.
Bowman's capsule	A cup-shaped receptacle in the vertebrate kidney that is the initial, expanded segment of the nephron where filtrate enters from the blood.
Boyle's law	It states that volume of a gas varies inversely with pressure.
Breeder seed	Breeder's seed is directly controlled by the originating or sponsoring plant breeding institution, firm or individual, and is the source for the production of seed of the certified classes. It has the maximum genetic purity.
Breeder's elevator	This euipment is used to lift the seeds to the top of bins/machines for cleaning, sorting or sacking.
Breeder's stock	A stock of higy pure seed of a variety in the custody of the plant breeder or plant breeding institution which orginated the variety.

Term	Definition
Breeder's stock seed	The seed harvested from a nucleus or a breeder's stock seed field.
Brownian Movement	Rapid oscillation motion of pollen grains in a drop of water is known as Brownian movement
Bucket elevator	This equipment is used to lift the seeds to the top of bins/machines for cleaning, sorting or sacking.
Buffer	A substance that consists of acid and base forms in solution and that minimizes changes in pH when extraneous acids or bases are added to the solution.
Buffer	A substance or group of substances that resist a change in hydrogen-ion concentration of a solution.
Buffer	A solution containing a mixture of weak acid and a base, which resists changes in pH and is therefore able to provide a favourable environment for enzymatic reactions.
Bulb	An enlarged, fleshy, thick, undergound part of a stem surrounded by a mass of leafy scales. Scales of a bulb are actually thickened and shortened leaves. Roots develop from the base of a bulb, *e.g.,* onion.
Bulk	Assembly of similar materials from a collection into one or more bulks.
C.F.M. per SQ. FT.	Used for specifying airflow in cu ft per minute on the basis of floor area of 1 sq ft obtained by dividing the c.f.m. of air by the area of the floor in sq ft.
C/N ratio	The ratio of carbon to nitrogen in a feedstock, substrate or culture medium. Various systems require a C/N ratio that falls within certain specific limits I order to optimize the process and ensure its reliability.
C_3 plant	A plant that uses the Calvin cycle for the initial steps that incorporate CO_2 in to organic material, forming a three carbon compound as the first stable intermediate.

Term	Definition
C_4 plant	A plant that prefers the Calvin cycle with reactions that incorporate CO_2 into four-carbon compounds, the end product of which supplies CO_2 for the Calvin cycle.
Calmodulin	A calcium-binding regulatory protein residing in the cytoplasm and affecting certain enzymes.
Calorie	The amount of heat energy required to raise the temperature of 1 g of water 1°C; the amount of heat energy that 1 g of water releases when it cools by 1°C. The Calorie (with a capital C), usually used to indicate the energy content of food, is a kilocalorie.
Calorie	A unit in which heat energy is measured.
Calvin cycle	The second of two major stages in photosynthesis (following the light reactions), involving atmospheric CO_2 fixation and reduction of the fixed carbon into carbohydrate.
CAM plant	A plant that uses crassulacean acid metabolism, an adaptation for photosynthesis in arid conditions, first discovered in the family Crassulaceae. Carbon dioxide entering open stomata during the night is converted into organic acids, which release CO_2 for the Calvin cycle during the day, when stomata are closed.
Capsid	The protein shell that encloses the viral genome; rod-shaped, polyhedral, or more completely shaped.
Carbohydrate	A sugar (monosaccharide) or one of its dimmers (disaccharides) or polymers (polysaccharides).
Carbon fixation	The initial incorporation of carbon dioxide into organic compounds.
Carbonyl group	A functional group present in aldehydes and ketones, consisting of a carbon double-bonded to an oxygen atom.
Carboxyl group	A functional group present in organic acids, consisting of a single carbon atom double-bonded to an oxygen atom and also bonded to a hydroxyl group.

Term	Definition
Carboxylase	An enzyme that catalyzes a reaction in which carbon dioxide is incorporated into another molecule.
Carcinogen	Any cancer-inducing substance.
Carnivore	An animal, such as a shark, hawk or spider that eats other animals.
Carrier protein	A transport protein involved in facilitated diffusion, possessing a specific binding site for a specific substance.
Cartilage	A type of flexible connective tissue with an abundance of collagenous fibers embedded in chondrin.
Caryopsis	A fruit developed from a single carpel with pericarp united to seeds; the fruit of cereals and grasses.
Catabolic pathway	A metabolic pathway that releases energy by breaking down complex molecules into simpler compounds.
Catabolic reaction	The reactions by which substrates are degraded to simpler substances in the cell
Catabolism	Degradative reactions in metabolism.
Catabolite activator protein (CAP)	In *E. coli*, a helper protein that stimulates gene expression by binding within the promoter region of an operaon and enhancing the promoter's ability to associate with RNA polymerase.
Catalase	An enzyme of wide distribution in plants and animals, which accelerates the decomposition of hydrogen peroxide with the liberation of water and free oxygen.
Catalyst	A substance that change the rate of a chemical reaction, usually accelerating it. The catalyst is not consumed in the process, nor does it affect the equilibrium constant of the reaction.
Cation	An ion with a positive charge, produced by the loss of one or more electrons.

Term	Definition
Cation exchange	A process in which positively charged minerals are made available to plant when hydrogen ions in the soil displace mineral ions from the clay particles.
cDNA clone	A selected host cell with a vector containing a cDNA molecule from another organism.
Cell	The basic unit of life. The smallest unit capable of independent reproduction.
Cell adhesion molecules (CAMs)	A diverse group of molecules on the surface of cells that contribute to selective cell association during embryonic development.
Cell wall	Unique to plant cells, a wall formed of cellulose fibers embedded in a polysaccharide-protein matrix. The primary cell wall is thin and flexible, whereas the secondary cell wall is stronger and more rigid and the primary constituent of wood.
Cellular respiration	The most prevalent and efficient catabolic pathway for the production of ATP, in which oxygen is consumed as a reactant along with the organic fuel.
Cellulose	A structural polysaccharide of cell walls, consisting of glucose monomers joined by β–1, 4-glycosidic linkages.
Central nervous system (CNS)	In vertebrate animals, the brain and spinal cord.
Central seed committee	The Central Seed Committee was constituted under sub-section (1) of section 3 of the Indian Seeds Act, 1966. Its function is to advice the Central Government and the State Governemnts on matters arising out of the administration of the Seeds Act, 1966 and to carry out the other functions assigned to it by or under the Seeds Act, 1966.
Central seed laboratory	The Central Seed Laboratory was established, or declared as such under sub-section (1) of Section 4 of the Seeds Act, 1966. The Seed Testing Laboratory at I.A.R.I., New Delhi has been notified as the Central Seed Testing Laboratory under the Act mentioned above.

Term	Definition
Centrifugal divider (gamet type)	This equipment is also commonly used for dividing a sample in the seed-testing laboratory. It makes use of centrifugal force to mix and scatter the seeds over the dividing surface.
Cephalin	A phospholipid, similar to lecithin, present in the brain of mammals.
Cerebroside	A lipid or fatty substance present in nerves and other tissues.
Certification (seed)	A system of maintaining the quality of seeds. The crops offered for certification are raised as per requirements for seed certification established by a seed certification agency. Several inspections are made to ensure purity and quality of seed.
Certification agency	The certification agency established under Section 8 or recoganized under Section 18 of the Seeds Act, 1966. Its major function is to certify seeds of any notified kind or varieties.
Certification sample	A sample of seed drawn by a certification agency or by a duly authorized representative of a certification agency.
Certification tag	Means a tag or label of a specified design of the certification agency. It constitutes a certificate granted by the certification agency.
Certified seed	The progeny of breeder's select, foundation or registered seed, so handled as to maintain satifactory genetic purity and identity. Production must be acceptable to a certification agency. Also refers to seed that fulfils all requirements for certification provided by the Seeds Act and Rules and to the container of which the certification tag is attached.
Certified Seed	It is produced from foundation/registered seed by progressive farmers, Government farms, under supervision of State Seed Corporation (Blue coloured certificate tag).
Certified seed producer	Means a person who grows or distributes certified seeds in accordance with the procedure and standards of the certification agency.

Term	Definition
Character	An identifiable hereditary property such as colour, leaf type etc.
Charle's law	It states that the volume of a gas varies directly with the absolute temperature.
Check	A row or plot standard variety included in nursery or plot tests for comparison.
Chelation	A combination of metallic ions with certain heterocyclic ring structures so that the ion is held by chemical bonds from each of the participating rings.
Chemical equilibrium	The rates of the forward and reverse reactions become equal, so that the concentrations of reactants and products stop changing such type of mixture is said to be chemical equilibrium.
Chemical hybridizing agents	Chemicals, which cause pollen abortion and render the treated plants male-sterile while not affecting the ovule fertility.
Chemical potential	The free energy per mole of a chemical substance is called its chemical potential.
Chemiosmosis	The ability of certain membranes to use chemical energy to pump hydrogen ions and then harness the energy stored in the H^+ gradient to drive cellular work, including ATP synthesis.
Chemoautotroph	An organism that needs only carbon dioxide as carbon source but that obtains energy by oxidizing inorganic substances.
Chemoautotroph	An orgnism that obtains energy for growth by oxidation of reduced compounds of sulphur, hydrogen, ammonia or nitrite.
Chemoheterotrop	An organism that must consume organic molecules both for energy and carbon.
Chitin	A structural polysaccharide of an amino sugar found in many fungi and in the exoskeletons of all arthropods.
Chlorophyll	A green pigment located within the chloroplasts of plants; chlorophyll a can participate directly in the light reactions, which convert solar energy to chemical energy.

Term	Definition
Chloroplasts	Chlorophyll-containing structures found in the cytoplasm of green plant cells. Photosynthesis takes place in the chloroplasts.
Cholesterol	A steroid forming an essential component of animal cell membranes and acting as a precursor molecule for the synthesis of other biologically important steroids.
Cholesterol	A high-molecular-weight cyclic alcohol contained in food of animal origin and synthesized in the body from Acetyl-CoA.
Chromatography	It is a separation method in which mixtures are resolved by differential migration of their constituents during passage through a column.
Chromogene	A heredity determiner in the chromosome in contrast to determiners in the cytoplasm.
Chromosome	A filamentous body in the cell nucleus, which is conspicuous during certain stages of cell division. The chromosomes of the cell contain the genes; or A deep staining, rod like body in the nucleus of cells, visible at cell division. Chromosomes contain the genes, the hereditary determiners. All species have characteristic chromosome numbers.
Chromosomes	Small thread or rod shaped structure in the common ancestory, which have been propogated vegetatively.
Chylomicron	A small lipid droplet composed mainly of dietary triacylglycerols, fat-soluble vitamins, small amounts of cholesterol and phospholipids, and a thin coating of protein.
Climax	The final stage of vegetation development after the stabilization is called climax stage.
Clone	A lineage of genetically identical individuals.
Codon	A group of three bases in fact codes for one amino acid. This group of bases is called as codon. UAG, UAA and UGA are the only three codons that do not specify an amino acid. The genetic code is non-over lapping. Codons that specify the same amino acid are called synonyms; *e.g.,* CAU and CAC are synonymous for the histidine. Most synonymous differ only in the last base of the triplet.

Term	Definition
Coenzyme	A small organic molecule associated with the protein portion of a holoenzyme, weakly bound at the active site of the enzyme and required for enzyme activity.
Coenzymes	A small molecular weight substance required for the catalytic activity of one or group of enzymes; or an enzyme often contains a tightly bound small molecule, termed a coenzyme that is essential for the activity of the enzyme.
Cofactor	Synonymous with coenzyme. Often used to refer to a substance of unknown structure.
Cofactor	A small, inorganic or organic substance that is required for the activity of an enzyme.
Cofactor	It is a non-protein comound, which is essential for the catalytic activity. An enzyme-cofactor complex is termed as holoenzyme. The protein on its own is termed as apoenzyme.
Coleoptile	The forst leaf above the cotyledon, which encloses the stem tip and other leaves (sheath) or The sheath enclosing and protecting the apex of the axis of the embryo and young seedlings in certain monocotyledon.
Coleorhiza	The sheath, which surrounds the primary root in the embryo of grasses.
Coleorhiza or radicle sheath	Protective sheath/layers around the radicle.
Collagen	The principal structural protein of bones, teeth, skin, cartilage, tendons, cornea, and blood vessels.
Colloids	When particles of a substance are distributed throughout water in a stable manner the system is called colloidal.
Colour blindness	Inability to distinguish certain colours.
Committee	Same as Central Seed Committee.
Community	An aggregation of living organisms having mutual relationship among themselves and to the environment.

Term	Definition
Competitive inhibitor	An inhibitor of an enzymatic reaction whose inhibition can be abolished by high concentrations of substrate. In other words the maximum velocity is the same in the presence and absence of the inhibitor. This effect is usually interpreted in terms of a competition between the inhibitor and the substrate for the active site of the enzyme.
Complementary DNA (cDNA)	DNA that is identical to a native DNA containing a gene of interest except that the cDNA lacks noncoding regions (introns) because it is synthesized in the laboratory using mRNA templates.
Complete protein	A protein food that contains all the essential amino acids in relatively the same proportions, as humans require.
Complete proteins	Proteins that contain ample amounts of all nine/eight essential amino acids.
Complete record	The information which relates to the origin, variety, kind, germination and prity of seeds of any notified kind or variety, offered for sale, sold or otherwise supplied.
Composite sample	A sample formed by combining and mixing all the primary samples taken from the lot.
Composite variety	A composite variety is developed by intermating (3-4 generations) selected open pollinated parent varieties.
Conditioning and pre-cleaning	This involves removing awns or hulls, breading up clusters, scalping off large trash and other operations, which improve the condition and flowability of the seed.
Conducting tissue	The principal water and food transporting tissue in a plant.
Conformation	The three dimensional arrangement of a molecule which depends upon the spatial orientation of its chemical bonds.
Conical divider	This equipment is commonly used for dividing samples in the seed-testing laboratory.

Term	Definition
Conjugation	A recombination mechanism that results in the transfer of genetic material between two bacterial cells that are temporarily joined.
Constitutive enzymes	Enzymes synthesized in fixed amounts independent of need.
Container	A box, tin, wrapper bags etc., used to pack any materials.
Contaminants	Th factors which affect the genetic and physical quality of seed such as off-types, foreign pollen, other crop plants, weed plants, plants affected by designated diseases weed seeds etc., are referred to as contaminants.
Continuous drying	Drying carried out with a continous flow of grain and air, in contrast to batch operation.
Contraception	The prevention of pregnancy.
Convection	The mass movement of warmed air or liquid to or from the surface of a body or object.
Conveyors	The equipment used for moving the seeds.
Cori cycle	During vigorous exercise, pyruvate produced by glycolysis in muscle is converted to lactate-by-lactate dehydrogenase. The lactate diffuses into the blood stream and is carried to the liver. Here it is converted to glucose by gluconeogenesis. The glucose is released into the bloodstream and becomes available for uptake by muscle (as well as other tissues, including brain). This cycle of reactions is called the cori cycle.

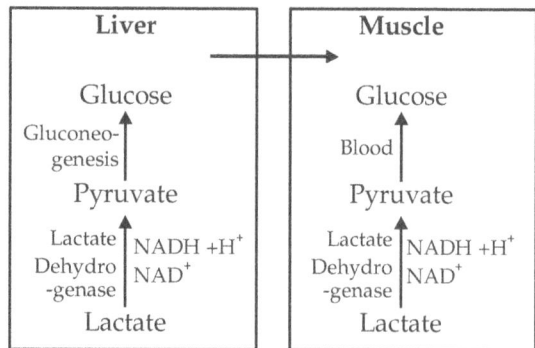

Term	Definition
Cori cycle	The process by which lactate produced in skeletal muscles during contraction is recycled in the liver to glucose.
Corm	A bulky, short, vertical underground stem which stores food.
Cortex	The region of the root between the stele and epidermis filled with growth tissue.
Cotyledon	The first leaves of the embryo, one in mono-cotyledons, two or more in dicotyledons.
Cotyledons	The one (monocot) or two (dicot) seed leaves of an angiosperm embryo.
Coupled reactions	Two chemical reactions which have a common intermediate
Covalent bond	A chemical bond formed between two atoms by the sharing of a pair of electrons.
Crassulacean acid metabolism (CAM)	Higher plants in which the photosynthetic carbon reduction cycle is supplemented by a second carbon assimilation pathway based on an initial fixation of carbon dioxide by the enzyme phosphenol pyruvate carboxylase.
Cross-fertilise	To fertilise the ovule or ovules of one flower with the pollen from another flower; commonly refers to the fertilizing of flowers of one plant by pollen from another plant.
Cross-pollinate	To apply pollen of one flower to the stigma of another; commonly refers to the pollinating of the flowers of one plant by pollen from another plant.
Crown	The persistent base of a tufted perennial herbaceous plant as found in grasses, asparagus etc.
Crude fiber	What remains of dietary fiber after acid and alkaline treatment? This consists of primary cellulose and lignin.
Cumulative effect	The action of two alleles of gene giving a more pronounced effect than one in the heterozygous condition.

Term	Definition
Curd	A mass of flower primordium meristems as found in monocotyledons, two or more in dicotyledons.
Cuticular transpiration	A small fraction of foliar transpiration also takes place from the general surface of the leaf through the cuticle and it is therefore, known as cuticle transpiration
Cyme	A centrifugal inflorescence on which the secondary or lateral branches continue to grow and extended beyond main axis.
Cytokinesis	The division of the cytoplasm to form two separate daughter cells immediately after mitosis.
Cytokinins	A class of related plant hormones that retard aging and act in concert with auxins to stimulate cell division influence the pathway of differentiation and control apical dominance.
Cytopasmic male sterility	A type of male sterility conditioned by the cytoplasm rather than by nuclear genes and transmitted only through the female parent.
Cytoplasm	The region of a cell inside the cell membrane and outside the nucleus; or the protoplasm of the cell surrounding the nucleus.
Cytoplasm	Main contents of a cell in which the nucleus and other bodies are located.
Cytoplasmic genetic male sterility	In this type, male sterility is dependent upon the action of genes carried in the nucleus with a particular cytoplasm.
Cytoplasmic male sterility	A type of pollen sterility transmitted through the cytoplasm, maternally inherited.
Cytosine	Nitrogen base one of two pyrimidines found in both DNA and RNA.
Cytosol	The cytosol is the soluble part of the cytoplasm where a large number of metabolic reactions take place.
Dalton	The atomic mass unit. A measure of mass for atoms and subatomic particles.

Term	Definition
Dalton's law of partial pressure	The law states that in a mixture of gases, each gas exerts the same pressure as it would were to occupy the same volume alone.
Damping off	The collapse of seedlings, ascribed to the attacks of such fungus organisms as *Botrytis vulgaris* and *Pythium* species.
Dark reaction	That portion of the photosynthetic process, which does not require light. Chemical energy in the form of ATP and reduced ferrdoxin is produced from light energy in the light reaction, and this chemical energy is used to use to carry on biosynthesis in the dark reaction.
Dark reaction	The carbon assimilation through the photosynthetic carbon reduction cycle (PCR) and associated pathways, including the C_4 cycle and C_2 photorespiratory cycle.
Day length	The number of hours of light in each twenty-four hour cycle.
De novo synthesis	Refers to the complete synthesis of a substance from its ultimate precursors as opposed to the final steps in the synthesis.
Dead seeds	Seeds, which, at the end of test period are neither hard nor fresh and have not produced seedlings, are classified as dead seeds.
Debearder	Debearder is a preconditioning machine used to debeard barley, clip oats, break up lecerne seed pods, break up grass seed doubles, separate and polish vegetable seeds, decorticate sugarbeet seeds and remove extra glumes and hulls.
Decay	Breakdown of organic tissue usually associated with the presence of microorganisms.
Decortication	Removal of the pith and bark from fibrous and other tissues.
Defoliants	A chemical or method of treatment that causes only the leaves of a plant to fall off or abscise. The fruits remain attached.

Term	Definition
Degeneracy of the genetic code	One possibility is that degeneracy minimizes the deleterious effects of mutations. Degeneracy of the code may also be significant in permitting DNA base composition to vary over a wide range without altering the amino acid sequence of the proteins encoded by the DNA.
Degeneration	The progressive decrease in vigour of successive generations of plants, usually caused by unfavourable growing conditions or diseases.
Denaturation	A process in which a protein unravels and loses its native conformation, thereby becoming biologically inactive. Denaturation occurs under extreme conditions of pH, salt concentration and temperature.
Density	The number of individuals per unit area or volume.
Deoxyribonucleic acid (DNA)	A double-stranded helical nucleic acid molecule capable of replicating and determining the inherited structure of cell's proteins.
Depth factor	A depth which would contain enough grain so that if all the theoretical heat available for drying could be used, it would all dry to equilibrium in a period of tieme equal to the time of one-half response.
Desiccate	To dry thoroughly to remove moisture from an object to definitely below the normal level.
Designated diseases	Refers to the diseases specified for certification of seed and in whose regard cerification standards must be met.
Designated weeds	Refers to weed species specified for certification of seeds of a given kind and in whose regard certification standards must be met.
Desitometer	A device used to such chromatographic plates or electrophoretic gels to evaluate the amount of various compounds present on the plate or gels.
Detassel	To remove the tassel at the top of maize plant before pollen is released.

Term	Definition
Determination	The progressive restriction of developmental potential, causing the possible fate of each cell to become more limited as the embryo develops.
Devernalization	A process where stimulus to flowering received by treatment with low temperature, (i.e., vernalization) can be reversed.
Dew point temperature	It is the temperature at which misture condenses on a surface.
Dextran	Dextran consists of glucose residues linked mainly by $\alpha 1 \rightarrow 6$ bonds but with occasional branch points that may be formed by $\alpha\, 1 \rightarrow 2$, $\alpha\, 2 \rightarrow\, 3$ or $\rightarrow 1 \rightarrow 4$ bonds.
Diabetes mellitus	A disease in which the blood glucose concentration is unstable due to a relative or complete deficiency of insulin and a relative or great excess of glucagons.
Diaphanoscope	A dark box or purity workboard having a small opening in its top, through which a bean of light passes from below; used principally for determining the presence or absence of caryopses within the glumes in certain grass seed samples.
Dicotyledons	A group of plants so classified because the embryo usually has two cotyledons.
Dietary fiber	Substances in food (essentially from plants) that are not digested by the processes present in the stomach and small intestine.
Diffusion	The spontaneous tendency of a substance to move down its concentration gradient from a more concentrated to a less concentrated area.
Diffusion	The movement of the molecules of gases, liquids or solutes from the regions of higher concentration to region of lower concentration until the molecules are evenly distributed throughout the available space is known as diffusion.
Diffusion	Net movement of molecules from region of greater concentration to region of lower concentration of concerned molecules.

Term	Definition
Diffusion pressure deficit	The amount, by which the diffusion pressure of a solution is lower than that of its solvent at the same temperature and atmospheric pressure is called the diffusion pressure deficit (DPD).
Digestion	The process of breaking down food into molecules small enough for the body to absorb.
	The enzymatic hydrolysis of food that occurs in the gastrointestinal tract.
Dihybrid	A hybrid for two different genes.
Dioecious	Having stamens and pistils on different palnts.
Diploid	Having a chromosome number just twice the haploid gametic number.
Diploid	Organisms or cell with two sets of chromosomes.
Disaccharide	A sugar, which yields two monomer sugar units (monosaccharides) on hydrolysis or two units of sugar united by glucosidic bond.
Disc separator	It is an upgrading machine, which makes the separations on the basis of length differences. The disc separator basically consists of cast iron discs, which revolve together on a horizontal shaft inside a cylindrical body.
Discolouration	Alteration or loss of colour.
Diseased	Showing the effect of the presence of disease organism or abnormal physiological acitivty.
Dispersion	The distribution of individuals within geographical population boundaries.
DNA (deoxyribonucleic acid)	The purine and pyrimidine bases of DNA carry genetic information whereas the sugar and phosphate groups perform a structural role.
	The molecule that carries, in the sequence of its bases, the genetic language of a species or an individual.
DNA double helix	In a DNA double helix, the two strand of DNA are wound round each other with the bases on the inside and the sugar-phosphate backbones on the outside. The two DNA chains are held together by hydrogen bonds between pairs of bases; adenine (A) always pairs with thiamine (T) and guanine (G) always pairs with cytosine (C).

Term	Definition
DNA libraries	Genomic DNA libraries are made from the genomic DNA of an organism. A complete genomic DNA library contains all of the nuclear DNA sequences of that organism. A cDNA library is made using complementary DNA (cDNA) synthesized from mRNA reverse transcriptase. It contains only those sequences that are expressed as mRNA in the tissue or organism of origin.
DNA sequence	The DNA sequence is the sequence of A, C, G, and T along the DNA molecule, which carries the genetic information.
DNA synthesis	Studies have shown that RNA synthesis is essential for the initiation of DNA synthesis. Furthermore, a short stretch of RNA is covalently linked to newly synthesized DNA fragments. Thus, RNA primes the synthesis of DNA. $$DNA \xrightarrow{\text{transcription}} RNA \xrightarrow{\text{translation}} Protein$$
DNA-DNA hybridization	The comparison of whole genomes of two species by estimating the extent of hydrogen bonding that occurs between single-stranded DNA obtained from the two species.
DNA-ligase	A linking enzyme essential for DNA replication catalyzes the covalent bonding of the 3′ end of a new DNA fragment of the 5′ end of a growing chain.
DNA-polymerase	An enzyme that catalyzes the elongation of new DNA at a replication fork in the 5′ → 3′ direction by the addition of nucleotides to the existing chain.
DNA-probe	A chemically synthesized, radioactively labeled segment of nucleic acid used to find a gene of interest by hydrogen-bonding to a complementary sequence.
Domain	1. A structural and functional portion of a polypeptide that may be coded for by a specific exon; a globular region of a protein with tertiary structure. 2. A taxonomic category above the kingdom level; the three domains are archaebacteria, eubacteria and eukaryotes.

Term	Definition
Dominant factor	A heredity factor or gene possessed by one parent of a hybrid which causes a character to be manifested in the hybrid to the apparent or near excusion of the contrasted character in the other parent.
Dormancy	An internal condition of the chemistry or stage of development of a viable seed that prevents its germination although good growing temperature and moisture are provided. A state of dormancy is a resulting state that must be broken by time or special conditions before a seed will germinate at temperature and moisture levels suitable for growth.
Double cross	A double cross is a first generation hybrid between two single crosses.
Double top cross	The first generation hybrid between an approved single cross and an approved open-pollinated variety.
Draper separation	A seed processing machine, which separates seed their relative ability to roll or slide, which is in turn determined by shape and surface texture.
Drum mixer	A simple mixer commonly used for dry treatment of seeds. It consists of a drum; through which a pipe is run at an angle mounted on two sawhorses.
Dry bulb temperature	The temperature of the air as measured by an ordinary thermometer.
Dryer (Drier)	A unit, which provides the conditions for removing moisture, generally by forced ventilation with or without the addition of heat.
Drying	The removal of moisture to a point at which the rate of deterioration from chemical and biological activity is slowed down.
Drying front	In a bin of drying grain the advancing zone that includes all of the grain that has just reached some arbitrarily initial moisture content, the zone between the initial and final moisture content.

Term	Definition
Dt gene maize	A "genic mutagen" in chromosome 9 in maize causing a (chromosome 3) to mutate to A, yielding dotted kernels or leaves streaked with anthocyanin.
Dusts	Seed treating products formulated in the form of dusts.
Dyes	Dyes are sometimes used along with seed treatment. It serves three purposes: 1. as a visible means of evaluating the completeness of treatment coverage; 2. a warning that the seed has been treated; and 3. serves as colour brand of a seed company.
Dynamic state	A condition within a cell, tissue or organ in which a structure is constantly being produced and destroyed in such a way that there is no net gain or loss.
Ear	A large dense or heavy spike or spike-like infloresence, as the ear of maize. Also used for the spike-like panicle of such grasses as wheat, barley etc.
Ecology	"Ecology is the study of the reciprocal relationship between living organisms and their environment" Earnest Haeckel.
Ecosystem	Ecosystem is the basic functional unit of organisms and their environment interacting with each other and with their own components.
Electrochemical gradient	The diffusion gradient of an ion, representing a type of potential energy that accounts for both the concentration difference of the ion across a membrane and its tendency to move relative to the membrane potential.
Electrogenic pump	An ion transport protein generating voltage across the membrane.
Electrolysis	The decomposition of a compund by an electrical current or the destruction of tumours, hair roots etc., in a similar manner.

Term	Definition
Electron acceptor	A member of an oxidation/reduction pair that can receive electrons from the electron donor; thus it is an oxidizing agent or oxidant. Oxygen is the ultimate acceptor in the respiratory chain.
Electron carriers	NADH, NADPH, $FADPH_2$ are the major electron carrier or A substance which can gain or lose electrons reversibly and which participates in the transfer of electrons from one molecule to another.
Electron donor	A member of an oxidation/reduction pair that can donate electrons to an electron acceptor; thus it is an reducing agent or reductant. Hydrogen is a strong reducing agent whereas water is a weak reducing agent.
Electron microscopy	A technique for visualizing material at very high magnification with resolutions attainable down to about 10 Å. Beams of electrons rather than light rays are employed.
Electron transport chain.	The series of carriers responsible for transporting electrons from substrate to oxygen during respiration.
Electron volt	Energy gained by an electron passing through a potential of one volt.
Electronic colour sorter	The machine used for separating seeds by differences in colour. The electronic colour sorter views each seed individually with photo-electric cells. The seed is compared with a selected background, or range according to its colour. If its colour or shade falls within the reject range a blast of compressed air deflects the seed and sends it into the reject discharge spout.
Electrostatic seed separator	This machine separates the seed of different electrical properties essentially independent of difference in size, shape, weight or surface texture. It separates seed by allowing them to fall free in space and then using an electric field to detect some seed from their normal flight path.
Elevators	The equipment used for moving seeds.
Emasculation	Removal of stamens before they burst and shed their pollen.

Term	Definition
Embeyo	The rudimentary plant within the seed.
Embryo dormancy	Due to physiological immaturity of embryo.
Emulsifier	A compound that assits in the formation and stabilization of an emulsion.
Emulsion	When two immiscible liquids mixed together the solution are formed is known as emulsion.
Encrushed seed	Units more or less retaining the shape of the seed with the size and weight changed to a greater or lesser extent. The encrusting material may contain pesticides, dyes or other additives.
Endergonic process	A process, which proceeds with an increase in free energy. It is not spontaneous and must be driven by coupling to some other process, which can supply energy.
Endergonic reaction	A no spontaneous chemical reaction in which free energy is absorbed from the surroundings.
Endocytosis	Endocytosis is the uptake of macromolecules from the extra cellular space into the cell across the plasma membrane via the formation of an intracellular vesicle pinching off from the plasma membrane.
Endocytosis	Cell membrane encircles particle and brings it into cell by forming vacuole around it.
Endoplasmic reticulum (ER)	An extensive membranous network in eukaryotic cells, continuous with the outer nuclear membrane and composed of ribosome-studded (rough) and ribosome-free (smooth) regions.
Endosperm	The tissue of seeds developing from fertilization of the polar nuclei of the ovule by a second male nucleus that nourishes the embryo.
Endosperm	A triploid primary endosperm nucleus produces endosperm in triple fusion in the process of fertilization. Endosperm is the tissue which provides nutrition to developing embryo, which acts as food reservoir.
Energetic	The physical laws of energy and energy transformations that operate during every chemical reaction.

Term	Definition
Energy	The capacity to do work by moving matter against an opposing force.
Energy of fat	A gram of nearly anhydrous fat store more than six times as much energy as a gram of hydrated glycogen.
Energy production or consumption	$NADH + H^+ + \frac{1}{2} O_2 \leftrightarrow NAD^+ + H_2O$ $\Delta G^0 = -52.6$ Kcal/mol (exergonic), $\Delta E^0 = +1.14$ Volts $ADP + Pi + H^+ \leftrightarrow ATP + H_2O$ $\Delta G^0 = +7.3$ Kcal/mol (endergonic), $\Delta E^0 = -30.5$ KJ/mol. $ATP + H_2O \leftrightarrow ADP + Pi$
Energy used	The synthesis of one molecule of glucose from two molecules of pyruvate requires six molecules of ATP.
Energy value	The span of the respiratory chain is 1.14 Volts, which is corresponds to 53 Kcal.
Energy yield	Two ATPs are used in glycolysis and four ATPs are synthesized for each molecule of glucose so that the net yield is two ATPs per glucose. Under aerobic conditions, the two NADH molecules arising from glycolysis also yield energy via oxidative phosphorylation.
Energy yield from citric acid cycle	For each turn of the cycle, 12ATP molecules are produced, one directly from the cycle and 11 from re-oxidation of the three NADH and one $FADH_2$ molecules produced by the cycle by oxidative phosphorylation.
Energy yield from fatty acids	Complete degradation of palmitate (C16:0) in β-oxidation generates 35 ATP molecules from oxidation of the NADH andFADH$_2$ produced directly and 96 ATPs from the break down of the acetyl CoA molecules in the citric acid cycle. However two ATP equivalents are required to activate the palmitate to its acyl CoA derivative prior to oxidation. Thus the net yield is 129 ATPs.
Enhancer	A DNA sequence that recognizes certain transcription factors that can stimulate transcription of nearby genes.

Term	Definition
Enthalpy	The thermodynamic function of a system equivalent to the internal energy plus the product of the pressure and the volume.
Entropy	A quantitative measure of disorder or randomness, symbolized by S.
Entropy	The randomness or disorder of a system. A measure of the capacity of a system to undergo spontaneous change.
Environment	Environment being the aggregate of all external conditions and influences which affect the life and development of organisms at a given spot.
Enzyme	A highly specialized protein molecule that serves as a catalyst for a specific chemical reaction and is not consumed or permanently altered by the reaction.
Enzyme inhibitor	A substance that reacts to modify the conformation of the active center or the substrate-binding site of an enzyme, slowing or preventing its catalytic action.
Enzymes	Protein molecules, which are specific and efficient catalysts of certain chemical reactions; or a complex organic substance (a specific protein or biocatalyst) that accelerates (catalyzes) a specific chemical reaction; or protein catalysts called enzymes are mediators of the dynamic events of life; an enzyme catalyzes almost every chemical reaction in a cell.
Epicotyl	The growing point of the embryo, which gives rise to the shoot or the above ground part of the plant.
Epigeal	Plants in which the cotyledons appear above the surface of the soil.
Epigeal seed germination	When cotyledons pushed above soil surface due to rapid elongation of hypocotyls (portion of embryo below cotyledons).
Epigial germination	A type of germination in which cotyledons and shoot are carried above soil level by the elongating hypocotyls.

Term	Definition
Epistatic gene	A gene, which suppresses the action of another gene not at the same locus in the chromosome.
Equilibrium	A state in which there are no further changes in the properties of a system with time. For a chemical reaction this means that the concentrations of reactants and products do not change further with time.
Equilibrium moisture content	The moisture content of the product when it is in equilibrium with the surrounding atmosphere is called the equilibrium moisture content or hydroscopic equilibrium.
Equilibrium potential	The membrane potential for a given ion at which the voltage exactly balances the chemical diffusion gradient for that ion.
Erg	Work performed by a force of one dyne moving its point of application one cm.
Erythrocyte	A red blood cell contains hemoglobin, which functions in transporting oxygen in the circulatory system.
Essential amino acids (indispensable)	Amino acids not efficiently synthesized by humans that must therefore be included in the diet. There are eight amino acids.
Essential nutrient	1. A chemical element required for a plant to grow from a seed and complete the life cycle. 2. A nutrient substance that an animal cannot make itself from raw materials but that must be obtained in food in prefabricated form.
Ester	The product of a reaction of an acid with an alcohol in which water is removed.
Ethylene	The only gaseous plant hormone, responsible for fruit ripening, growth inhibition, leaf abscission, and aging.
Eukaryotes	Eukaryotic cells have a membrane-bound nucleus and a number of other membrane-bound sub cellular (internal) organelles, each of which has a specific function.

Term	Definition
Eukaryotes	Organisms whose cells contain a membrane bounded nucleus. With the exception of the bacteria and blue-green algae, all living organisms are eukaryotes.
Evolution	The process by which gradual shifts in the nature of a living population leads, after many generations, to the appearance of new characteristics in the population.
Exergonic process	A process, which proceeds with a decrease in free energy. A spontaneous process.
Exergonic reaction	A spontaneous chemical reaction in which there is a net release of free energy.
Exocytosis	Exocytosis is the secretion of proteins out of the cell across the plasma membrane into the extra cellular space.
Exocytosis	Cell membrane ejects materials; vesicle filled with material fuses with cell membrane.
Exoenzyme	An enzyme localized on the outside of the cell membrane.
Exon	The coding region of a eukaryotic gene that is expressed. Exons are separated from each other by introns.
Exotoxin	A toxic protein secreted by a bacterial cell that produces specific symptoms even in the absence of the bacterium.
F_1	Denotes the first generation offspring coming from the mating of two parents.
F_1 generation	The first filial or hybrid offspring in a genetic cross-fertilization.
F_1 hybrid	The first generation resulting from a crossmating of distinctly different parental types.
F_2	The second filial generation. The generation produced by selfing the F_1.
F_2 generation	Offspring resulting from interbreeding of the hybrid F_1 generation.

Term	Definition
Facilitated diffusion	Carrier protein in cell membrane accelerates movement of relativity large molecules from region of their higher to region of their lower conentration.
Faraday	The change in one mole of electrons to 96,500 coulombs.
Farnesol	A colourless liquid extracted from oils of plant such as citronella, neroli, cyclamen and tuberose. It has delicate floral odour.
Fat	A biological compound consisting of three fatty acids linked to one glycerol molecule.
Fatty acid	A long carbon chain carboxylic acid. Fatty acids vary in length and in the number and location of double bonds; three fatty acids linked to a glycerol molecule from fat.
Fatty acid synthesis	Fatty acid synthesis involves the condensation of two-carbon units, in the form of acetyl CoA, to form long hydrocarbon chains in a series of reactions. These reactions are carried out on the fatty acid synthase complex using NADPH as reductant. The fatty acids are covalently linked to acyl carrier protein (ACP) during their synthesis.
Favism	A hereditary condition common to persons native to the Mediterranean area resulting in sensitivity to a species of beans, *Vicia faba*.
Feedback inhibition	A method of metabolic control in which the end product of a metabolic pathway acts as an inhibitor of an enzyme within that pathway.
Fermentation	The biochemical degradation of sugar and other food stuffs by reactions which take place under anaerobic conditions (*i.e.*, in the absence of oxygen) OR A catabolic process that makes a limited amount of ATP from glucose without an electron transport chain and that produces a characteristic end-product, such as ethyl alcohol or lactic acid.
Fermentation	Chemical transformation induced by the activity of enzyme systems of microorganisms.

Term	Definition
Ferredoxin	It is an iron-sulphur protein. Ferredoxins have a very negative redox potential and are constituents of electron trnsport chains.
Fertility restorers	Cytoplasmic male sterile lines will produce viable pollen in certain genotypes. Restorers have genes that restore fertility to a cytoplasmic male sterile line.
Fertilization	The union of haploid gametes to produce a diploid zygote.
Fiber	A lignified cell type that reinforces the xylem of angiosperms and functions in mechanical support, a slender, tapered sclerenchyma cell that usually occurs in bundles.
Fiber	Polysaccharides for which humans do not have digestive enzymes; thus fiber contributes no calories to the human diet.
Fibrin	A protein synthesized from fibrinogen by the action of thrombin and required for blood clotting.
Fibrinogen	A blood protein used to synthesized fibrin, a protein necessary for blood clotting.
First generation hybrid	Same as F_1 hybrid.
First law of thermo-dynamics	The principle of conservation of energy. Energy can be transferred and transformed but it cannot be created or destroyed.
Five classes of immuno-globulin	Human immunoglobulins exist as IgA, IgD, IgE, IgG and IgM classes which contain α, δ, ε, γ, and μ heavy chains, espectively. IgM is a pentamer that binds to invading microorganisms and activates complement killing of the cells and phagocytosis. IgG is the main antibody found in the blood after antigen stimulation and also has the ability to cross the placenta. IgA mainly functions in body secretions. IgE provides immunity against some parasites but is also responsible for the clinical symptoms of allergic reactions. The role of IgD is unknown. All antibody molecules contain either Kappa (K) or Lambda (λ) light chains.

Term	Definition
Flail	A fairly heavy, flat piece of wood fastened to a length of rope or leather for use in threshing seed.
Flavoprotein	A protein molecule, which contains flavin adenine dinucleotide (FAD) or flavin mononucleotide (FMN) as a prosthetic group. The flavoprotiens are involved in many oxidation-reduction reactions in the cell.
Flocculation	When there is no any electric charge on the particle in the solution and that particles settle down in the solution this phenomenon is known as flocculation.
Floret	In grasses, the lemma and palea with included flower; stamens and pistil.
Fluorescence	Part of the radiant energy is re-emitted as light, which is called fluorescence.
Fluorescence	The property of substances that becomes luminous when exposed to ultraviolet and other forms of radiant energy.
Foliar transpiration	The transpiration, which takes place from the leaves, is known as foliar transpiration.
Foreign seeds	Seeds of weeds and crops other than the kind being tested.
Foundation seed	The progeny of breeder's slect or foundation seed handed to maintain specific genetic purity and identity. The production must be acceptable to a certifying agency. It is primary source of seed of a genetically identified variety from which all further increases are made.
	Progeny of breeder seed production by agencies of public sectors, *i.e.*, research stations, Government forms or private sectors under supervision of seed certification agency (white colour tag).
Foundation stock	Seed production by a foundation stock seed grower in conformity with the regulations for producing such seed and approved as foundation stock seed by the certification agency.
Free energy	The energy that is available to perform work as a biochemical system precedes towards equilibrium.

Term	Definition
Free energy change	A measure of the maximum useful work it is possible to obtain from a process carried out at constant temperature and pressure.
Free radical	Shortlived form of compounds that exist with an unpaired electron in their outer electron shell. This causes it to have an electron-seeking nature, which can be very destructive to electron-dense areas of a cell, such as DNA and cell membranes.
Free-flowing seed	Non-chaffy smooth seeds which will travel easily through sampling or mixing devices.
Fresh ungerminated seed	Seeds other than hard seeds, which remain firm and apparently viable, even after the appropriate treatment for breaking dormancy.
Full sib	Progeny from a cross between two slected plants within the population.
Fumigation	Fumigation of seed lots by celphos, E.D.B., etc. to kill storage insect pests. Storage structures must be leakproof for effective fumigation.
Function of cholesterol	Cholesterol is a component of cell membranes and is the precursor of steroid hormones and the bile salts.
G_1 phase	The first growth phase of the cell cycle, consisting portion of interphase before DNA synthesis begins.
G_2 phase	The second growth phase of the cell cycle, consisting portion of interphase after DNA synthesis occurs.
Galactosemia	A metabolic inability, inherited as a recessive trait, to convert galactose to glucose because of an absence of galactose 1-phosphate uridyl transferase.
Galactoside	A molecule containing galactose in which some group other than hydrogen is attached to the hydroxyl group at the C-1 atom of galactose. The prefix a or b refers to the orientation of this group above or below the galactose ring.

Term	Definition
Gall	Swelling or excrescence of the tissue of plants resulting from the attacks of certain parasites or seed structures in which the contents have been replaced by nematodes or celworms.
Gametophytic tissue	The nutritive tissue occurring within conifer seeds.
Gel-electrophoresis	Separation of nucleic acids or proteins on the basis of their size and electric charge by measuring their rate of movement through an electric field in a gel.
Gene	The unit of heredity. The factor controlling a hereditary trait; or One of many discrete units of hereditary information located on the chromosomes and consisting of DNA or the discrete particulate hereditary determiner located in the chromosome in linear order; the 'element' of Mendel and factor of early genetic terminology.
Gene amplification	The selective synthesis of DNA, which results in multiple copies of a single gene, thereby enhancing expression.
Gene cloning	Formation by a bacterium, carrying foreign genes in a recombinant plasmid of a clone of identical cells containing the replicated foreign genes.
Gene organization	Most protein-coding genes in eukaryotes consist of coding sequences called exons, interrupted by noncoding sequences called introns.
Generally recognized as safe (GRAS)	A group of food additives that in 1958 were considered safe, therefore allowing manufactures to use them thereafter when needed in food products.
Genetic code	The genetic code is the relationship between the sequence of bases in DNA (or its RNA transcripts) and the sequence of amino acids in proteins.
Genetic code is a triplet code	The genetic code is the set of rules that specify how the nucleotide sequence of an mRNA is translated into the amino acid sequence of a polypeptide. The nucleotide sequence is read as triplets called codons. The codons UAG, UGA, and UAA do not specify an amino acid and are called termination codons or stop codons. AUG codes for methionine and also acts as an initiation or start codon.

Term	Definition
Genetic engineering	Alteration of genetic material in plants or animals with the intent of improving growth, disease resistance or other characteristics.
Genetic purity	Trueness to type; varietal purity; palnts/seeds conforming to characteristics of the varity as described by the breeder.
Genetic shift	Change in the genetic make up of varieties, if grown over a long period in areas outside their adapatation.
Genetic sterility	A type of male sterility conditioned by nuclear genes in contrast to cytoplasmic sterility. Either the male or female parent may transmit it.
Genetics	The science of heredity of similarities and differences among organisms.
Genome	A set of chromosomes (n) inherited as unit.
Genomic clone	A selected host cell with a vector containing a fragment of genomic DNA from a different organism.
Genomic library	A set of thousand of DNA segments from a genome each carried by a plasmid or phase.
Genotype	The genetic constitution or gene makeup of an organism.
Genotype	The heredity make up of an individual plant or animal, which the environment controls the individual's characteristics such as type of flower or bony structure or shape of leaf or colur of hair.
Geotropism	Plant growth response to gravity.
Geotropism negative	Upward growth (normal shoot).
Geotropism positive	Downward growth (normally primary roots).
Germination	The resumption of growth by the embryo and development of a young plant from the seed. Germination in a laboratory test, it the emergence and development from the seed embryo of those essential structures which for the kind of seed being tested, indicate the ability to develop into a normal plant under favourable conditions in the soil.

Term	Definition
Germination cabinet	A type of seed germinator most commonly used in a seed laboratory.
Germinative	Having the ability to grow and develop.
Germplasm	A special kind of protoplasm transmitted unchanged from generation to generation.
Gibberellins	A class of related plant hormones that stimulate growth in the stem and leaves, trigger germination of seeds and breaking of bud dormancy and stimulate fruit development with auxin.
Gibberllins	These are the plant growth stimulating chemicals. These include many different plant effects such as rapid stem growth, overcoming of dormancy, production of seedlings fruits and other responses.
Glomerate	A compact cluster forming a round mass.
Glucagon	A polypeptide hormone secreted by the cells in the pancreas. Glucagon responds to low levels of blood glucose by activating cyclic AMP in the liver, thus stimulating gluconeogenesis and glyco-genolysis.
Gluconeogenesis	Gluconeogenesis synthesizes glucose from non-carbohydrate precursors and is important for the maintenance of blood glucose levels during starvation or during vigorous exercise. The brain and erythrocytes depend almost entirely on blood glucose as an energy source. Gluconeogenesis occurs mainly in the liver and to a lesser extent in the kidney. Most enzymes of gluconeogenesis are cytosolic, but pyruvate carboxylase and glucose-6-phosphate are located in the mitochondrial matrix and bound to the smooth endoplasmic reticulum respectively.
Gluconeogenesis	The synthesis of glucose from substances such as lactate, some amino acids, and glycerol. The gluconeogenic pathway converts pyruvate to glucose.
Glumes	The pairs of bracts at the base of a spikelet in grasses.

Term	Definition
Glycemic Index	A ratio used to measure the relative ability of a carbohydrate to raise blood glucose levels as opposed to the ability of white bread/glucose to raise blood glucose levels.
Glycogen	A polypeptide hormone that is secreted by the a-cells of the pancreas when the blood sugar level is low. This hormone increases the blood sugar level by stimulating the breakdown of glycogen in the liver.
Glycogen metabolism	Glycogen degradation and glycogen synthesis are controlled both by allosteric regulation and by hormonal control.
Glycogenesis	The synthesis of glycogen. Glucose produced from the digestion and metabolism of carbohydrate and protein foods that are eaten in excess of need may be stored in glycogen molecules in the liver and muscles.
Glycogenolysis	The chemical process by which glucose is freed from glycogen. Glucose is a vital source of energy, especially for neutral tissues, and the body has evolved an exquisite method for its storage as glycogen as well as a smooth-running process for reclaiming it from glycogen.
Glycolysis	A group of chemical reactions responsible for converting glucose to lactate and ATP without the consumption of oxygen, an anaerobic path. OR Glycolysis is a set of reactions that take place in the cytoplasm of prokaryotes and eukaryotes. The roles of glycolysis are to produce energy (both directly and by supplying substrate for the citric acid cycle and oxidative phosporylation) and to produce intermediates for biosynthetic pathways.
Gonad	A gland such as an ovary or testis in which gametes are produced.
Gout	A metabolic disease marked by an excess of uric acid in the blood accompanied by painful inflammation of the joints.
Graham's law	It states that relative speeds of diffusion of gases are inversely proportional to the square roots of relative densities.

Term	Definition
Grains of moisture	The unit of measurement used to determine the amount of moisture in the air. 7000 grains equal one pound of water.
Gravity separator	It is a seedprocessing machine by which seed separations are made on the basis of differences in specific gravity.
Green house effect	The warming of the earth due to atmospheric accumulation of carbon dioxide, which absorbs infrared radiation and slows its escape from the irradiated earth.
Grow-out test	Performed to determine the genuineness of seed as to species or variety or freedom from seed-borne infection.
Growth	An increase in the size or number of cells representing a net increase in total cellular matter.
Guanine	A nitrogen base one of two purine found in both DNA and RNA.
Half-cell reaction	A chemical reaction, which describes a substance gaining or losing electrons and which therefore, describes one half of an oxidation-reduction reaction.
Half-life	The time required for half of a substance to disappear from a system.
Haploid	Having half the number of chromosomes found in diploid organisms.
Hard seeds	Seeds of *Leguminosae* and *malvaceae*, which remain hard at the end of pprescribed test period because they have not absorbed water owing to an impermeable seed coat, are classified as hard seeds.
Hardy-Weinberg law	A law concerning gene frequencies in populations. It states that after one generation of random mating gene frequencies remain constant in future generations.
Heat injury	Te injury caused to seeds due to heat; drying at higher temperatures may cause heat injury to seeds and may result in the lowering of seed germination.

Term	Definition
Heme	The portion of hemoglobin, myglobin, and cytochromes that is responsible for their ability to carry and release oxygen and electrons.
Hemochromatosis	A rare disease of iron metabolism in which iron accumulations in body tissue. The liver becomes enlarged, the skin takes on a bronze hue, diabetes mellitus may develop, and cardiac failure is common.
Hemolysis	Rupture of red blood cells, resulting in anemia.
Henderson-Hasselbalch equation	An equation relating the pH, the pK, and the ratio of the concentrations of a weak base to that of a weak acid.
Henry's law	According to this law if there is no chemical reaction between a gas and a liquid, the amount of gas absorbed in the liquid will be proportionate to its partial pressure.
Hereditary	Transmittable from parent to offspring or progeny.
Heredity	The transmission of genes and there by traits controlled by these genes from parent to offspring.
Heritability	The extent to which a given trait is determined by inheritance.
Heritable	Capable of being passed by inheritance.
Heterogeneity	Refers to the variation in a seed lot.
Heteroploid	Having a chromosome number differing from the normal number.
Heterosis	Hybrid vigor a term coined by George H. Shull.
Heterozygous	Not true breeding for a specific hereditary character. Plants may be heterozygous for some characters and homozygous for others.
High density lipoprotein (HDL)	One of the lipoproteins in which lipids are carried in the ewatery fluids of the body.
High energy compounds	A substance whose hydrolysis results in the release of a large amount of energy.
Hilum	A scar left where the seed stalk of funiculus breaks amawy or where the seed was attached directly to the placenta when there is no seed stalk.

Term	Definition
Histidinemia	An accumulation of histidine in the blood and other tissues, resulting from the absence of histidase.
Histones	Histones play a role in the packing of DNA molecules, rendering them more compact. Some of the nonhistone proteins associated with chromosomes are more likely participants in the specific control of gene function.
HIV	Human immunodeficiency virus. The infections agent that causes AIDS; HIV is an RNA retrovirus. AID (Acquired immunodeficiency syndrome). The name of the late stages of HIV infection; defined by a specified reduction of T cells and the appearance of characteristic secondary infections.
Holoenzyme	The complete, active enzyme. The protein parts, called the apoenzyme, plus the catalytic part, called the cofactor or coenzyme, make up the holoenzyme, which has biologic activity.
Homeostasis	A state of physiological equilibrium produced by a balance of functions and of chemical composition within an organism.
Homocystinuria	An inherited disease characterized by a high level of homocysteine excreted in the urine.
Homologous chromosomes	Chromosomes which pair at meiotic prophase and are similar in size, shape, structure and function. They have alleles of the same genes.
Homozygous	True breeding for a specific hereditary character. A plant that breeds true for a character such as flower colour is called homozygous for this character. Plants may be true breeding for some characters and not for others.
Horizontal or Flat storage	Storage structures the height of which is less then the width or diameter.
Hormone	A chemical substance produced by one tissue in an organism, which exerts some controlling influence on other tissues in the same organism; or One of many types of circulating chemical signals in all multicellular organisms that are formed in specialized cells, travel in body fluids and coordinate the various parts of the organism by interacting with target cells.

Term	Definition
Hormone	A chemical substance synthesized in one organ or gland that acts as a messenger to stimulate or inhibit the reactions in another organ or tissue. Hormones may be steroids, proteins or protein derivatives.
Host cell	A cell (usually a bacterium) in which a vector can be propagated.
Hybrid	A cross of unlike organisms.
Hybrid vigour	The increase in vigour over the parental types exhibited by hybrids.
Hybridization	The process of making a hybrid by cross-pollination in plants or by mating two types of animals.
Hydrogen bond	A weak attractive force between a hydrogen atom and a second atom. Such bonds are generally formed only if the second atom is very electronegative (such as N, O or Cl) and the hydrogen atom is also covalently bound to an electronegative atom.
Hydrogenation	The process of adding hydrogen atoms to a substance to alter its physical or chemical characteristics.
Hydrolysis	The cleavage of a chemical bond by reaction with water.
Hydrolysis	The process by which a molecule is broken apart with the addition of water.
Hydrophilic	Having an affinity for water
Hydrophilic	A descriptive term indicating that a molecule is polar (has regions of opposing charges) and associates with water molecules.
Hydrophobic	Having an aversion to water, tending to coalesce and form droplets in water.
Hydrophobic	A descriptive term indicating that a molecule is nonpolar and is repelled by water molecules.
Hydrophobic interactions	Nonpolar molecules contain neither ions nor dipolar bonds and thus these molecules do not become hydrated. Because they are insoluble or almost insoluble in water they are called hydrophobic.

Term	Definition
Hyperglycemia	High blood glucose levels above 140 milligrams per 100 ml of blood.
Hyperkalemia	A serum potassium level above 6 mEq/L. A level of 7 mEq/L is considered an emergency situation.
Hyperuricemia	An increase in the serum level of uric acid.
Hypocotyl	The part of the embryo or seedling below the cotyledonary node and above the root; the transition region connecting the stem and root.
Hypogeal	Plants in which the cotyledons remain below the surface of the soil.
Hypogeal germination	A type of germination in which the cotyledons or comparable structure remain in the soil and within the seed.
	When cotyledons remains below soil surface due to rapid elongation of epicotyl (portion of embryo above cotyledons).
	When cotyledons remains below soil surface due to rapid elongation of epicotyl (portion of embryo above cotyledons).
Hypoglycemia	Low blood glucose levels, below 40 to 50 mg/100 ml of blood.
Hypokalemia	A serum potassium level below 3.5 mEq/L.
Hypostatic gene	A gene whose phenotypic effects are suppressed by another gene not at the same locus in the chromosome.
Imbibition	Absorption of moisture by a colloidal substance such as seed coats which is accompanied by swelling of tissues.
Imbibition pressure	Matric potential is known as imbibition pressure.
Immature	Not fully developed not having parts developed.
Immunoprotein	A protein, usually found in the gamma globulin fraction of blood plasma, that develops in response to an antigen and reacts with the antigen to cause its destruction.

Term	Definition
Impermeable	Impenetrable as when a seed coat allows no passage through to water or gases.
***In Situ* hybridization**	For *in situ* hybridization a tissue sample is incubated with a labeled nucleic acid probe, excess probe is washed away and the location of hybridized probe is examined. The technique enables the spatial localization of gene expression to be determined as well as the location of individual genes on chromosomes.
Inbred line	An inbred line is a relatively true breeding strain resulting from at least five successive generations of controlled self-fertilisation or of back-crossing to a recurrent parent with selection or its equivalent.
Incipient wilting	Partial loss of turgidity does not cause visible wilting and is known as incipient wilting.
Inclined separator	A seed processing machine, which separates seed their relative ability to roll or slide, which is in turn determined by shape and surface texture.
Increase	To multiply a quantity of seed by palnting it rearing the palnts that grow from it, and harvesting the seeds they produce. The seeds resulting from this process are called an increase.
Incubation	Maintaining seeds in an equivalent favourable to development of pathogens or symptoms.
Indented cylinder separator	The imdented cylinder separator is a rotating almost horizontal, cylinder with a movable horizontal sparating trough mounted inside it. The length of the seed is the only factor affecting the separation made by the indented cylinder.
Indexing	Refers to the process used to test vegetatively reproduced plants for freedom from virus diseases before multiplying them.
Induced chromosome break	Chromosome break caused by some agent usually external to the chromosome, such as radiation or chemicals.

Term	Definition
Inert matter	Inert matter includes seeds and seed like structures, namely pieces of broken or damaged seeds which are one-half the original size or less; empty glumes, lemmas, paleas, unattached sterile florets and florets with a caropsis less than the minimum size; prescribed and other matter, namely soil, sand, stones, chaff, stems, leaves, cone scales, winges, pieces of bark, flowers nematode galls, fungus bodies, caropses of graminae replaced by insect larvae and all other matter not seeds.
Infected	Carrying a disease organism but not necessarily showing the symptoms of disease.
Inferior result	A result obtained in a subsequent test at another station of a second submitted sample from the same seed lot, which indicates a lower quality than the first test, *e.g.*, a lower purity percentage a lower germination capacity a higher count of certain seeds or a higher percentage of other seeds.
Infested	Carrying an insect pest, but not necessary showing the insect damage.
Inflorescence	The flowering axis or other specialized flowering structure of a plant such as an umbel, racem, spike, tassel and panicle.
Inhibitor	A substance, which lowers the rate of a chemical reaction. (Some times called a negative catalyst)
	A chemical substance that acts tyo prevents a process from occurring; many chemicals both natural and artifical can act to prevent seed germination.
Initial velocity	The rate of a chemical reaction at zero time when essentially no reactant has been converted to product.
Inoculum	Material such as spores, bacteria etc., used for infecting a plant with a disease or for propagating microorganisms in controlled cultures.
Inositol	An alcohol that is a part of a phosphoglyceride in cell membrane structures.
Inseparable other crop plants	Crop plants whose seed is difficult to separate from the main crop seed.

Term	Definition
Insulin	The vertebrate hormone that lowers blood sugar levels by promoting the uptake of glucose by most body cells and promoting the synthesis and storage of glycogen in the liver; also stimulates protein and fat synthesis; secreted by endoplasmic cells of the pancreas called is lets of Lange Hans.
Insulin	A hormone, secreted by the beta cells of the islets of Langerhans in the pancreas, that enhances the uptake of glucose by the peripheral tissues and thus maintains the blood glucose level within normal limits.
International seed analysis certificate	A form of certificate issued only by the International Seed Testing Association and used for reporting the results of tests.
Interpretation/Germination	A seed shall be cosidere to have germinated when it has developed into a normal seedling. Broken seedling and weak, malformed and obviously abnormal seedling shall not be considered to have germinated.
Intrinsic factor	A protein secreted by the gastric cells and required for the absorption of Vitamin B_{12}.
Inversion in genetics	A rearrangement of a chromosome in which a portion is rotated a full 180 degrees. It sometimes results when a chromosome is broken at two places.
Ion	An atom with an unequal number of electrons and protons. If the number of electrons exceeds the number of protons, the ion is negative. If the number of protons exceeds the number of electrons the ion is positive.
Ionizing radiation	Any of the high-energy radiations that displace electrons from neutral atoms.
Isoelectric focusing	In isoelectric focusing, proteins are separated by electrophoresis in a pH gradient in a gel. They separate on the basis of their relative content of positively and negatively charged residues. Each protein migrates through the gel unit it reaches the point where it has no net charge, its isoelectric point (PI).

Term	Definition
Isoelectric point	The pH of the solution at which a protein will posses an equal number of positively and negatively charged groups, *i.e.,* the protein posses on net charge.
Isoenzymes	Isoenzymes are different forms of an enzyme, which catalyze the same reaction, but which exhibit different physical or kinetic properties.
Isolation	Separation of seed fields from fields of other varieties of the same crop same variety fields not conforming to varietal purity requirements; other related species, fields and fields affected by designated diseases to prevent genetic and disease contamination.
Isolation distance	Distance to be maintained between the seed crop and the contaminant.
Isolation distance	The proper distance should be maintained to avoid the contamination through natural crossing. The separation of seed production plot from the same/other crop according to crop and stage or class of seed.
Isolation requirements	Refers to the isolation required for maintaining the desired purity and health of a crop, such as the minimum specified isolation required for seed certification.
Isomer	Different chemical structures for compounds that share the same chemical formula or one of several organic compounds with the same molecular formula but different structures and therefore different properties. The three types are structural isomers, geometric isomers and optional isomers.
Isoprene	A five carbon unit important in the synthesis of many biochemicals known for their fragrance, such as bay leaves or menthol, and other molecules some of which lend colour to tomatoes and carrots.
Isotope	An alternate form of a chemical element. It differs from other atoms of the same element in the number of neutrons in its nucleus.

Term	Definition
Jaundice	A yellow colour in the skin that is evidence of an accumulation of bilirubin, a bile pigment produced in the spleen during the breakdown of the heme portion of hemoglobin, myoglobin, and cytochromes.
Joule (electrical)	Energy developed when one coulomb of electrons (10.364×10^{-6} moles) passes through a potential of one volt.
K′	A constant that is the pH at which the protonated and unprotonated components are equal. Stated in other terms pK′ is the pH at which an acid is one-half dissociated. The pK′ of each acid is unique.
Karyotype	The character of the chromosomal complement with reference to the comparative size, shape and morphology of the different chromosomes.
Ketone bodies	When in excess, acetyl CoA produced from the β-oxidation of fatty acids is converted into acetate and D-3- hydroxybutarate. Together with acetone, these compounds are collectively termed ketone bodies. Acetoacetate and D-3- hydroxybutarate are produced in the liver and provide an alternative supply of fuel for the brain under starvation conditions or in diabetes, *e.g.*, Acetoacetate, β-hydroxybutyrate and acetone.
Ketonemia	A high level of ketone bodies in the blood.
Ketonuria	A high level of ketone bodies in the urine.
Ketosis	Metabolic acidosis caused by a rise in the plasma level of the acids, acetoacetate, β-hydroxybutyrate and acetone.
Kilo calories per Einstein	Energy of mol of photons (one Einstein) of wavelength in m: Kcal per mol = 28,589.7/m.
Kind	Means one or more related species or sbu-species of crop plants each individually or collectively known by one common name, *e.g.*, wheat, paddy, etc.
Kinetic energy	It is the energy of movement the motion of a car, for example, or the motion of molecules.

Term	Definition
Lag phase	The period between the initial in oculation of a culture medium with an organism and the resumption of normal growth.
Land requirement	Refers to the requirements as to preceding crop for seed certification.
Law of independent assortment	Mendel's second law, stating that each allele pair segregates independently during gamete formation; applies when gene for two traits are located on different pairs of homologous chromosomes.
Law of segregation	Mendel's first law, stating that allele pairs separate during gamete formation and then randomly reform pairs during the fusion of gametes at fertilization.
Lecithin	A group of phospholipids containing two fatty acids a phosphate group and a choline molecules.
Lecithin	The travel name of a phosphoglyceride whose correct name is phosphatidylcholine.
Lemma	The outer bract of the flower of grasses sometimes referred to as flowering glume.
Lenticular transpiration	Loss of water vapour also takes place through the lenticels of fruits and woody stems and is called lenticular transpiration.
Lesion	A wound, well-marked but limited diseased area.
Library (genomic)	A complete set of genomic clones from an organism or of cDNA clones from one cell type.
Ligand	A molecule that binds specifically to a macro-molecule (other than the substrate or products of an enzyme) is often called a ligand of that macromolecule.
Light reaction	That potion of the photosynthetic process in which light energy is converted into chemical energy.
Lignin	An insoluble fiber made up of a multiringed alcohol structure.
	A complex carbohydrate polymer making up about 25% of the wood of trees and also found in the cell walls of schlerenchyma tissues and vessels, fibers and tracheids at maturiy.

Term	Definition
Limiting amino acid	The essential amino acid in the lowest concentration in a food in comparison with the body's need.
Line	A group of individuals from a common ancestry; a more narrowly defined group than a strain or variety.
Lipid	One of a family of compounds including fats, phospholipids, and steroids that are insoluble in water.
Lipogenesis	The synthesis of lipids.
Lipolysis	The hydrolysis of lipids by enzymes termed lipases.
Lipoproteins	Conjugated proteins consisting of simple proteins combined with lipid components, the latter including cholesterol, phospholipids or triacylglycerols. (Lipoproteins and their density: Chylomicrons <0.960, VLDL, 0.960 – 1.006, LDL, 1.006 – 1.059, HDL, 1.059 – 1.210, VHDL, 1.210 g/cm^3).
Locus	The physical location of a gene in the chromosome.
Lot	A lot is a specified quantity of seed, which is physically identifieable.
Low input sustainable agriculture (LISA)	A from of farming that attempts to limit use of purchased materials such as manufactured fertilizers and pesticides. Use of manure and crop rotation is typical substitutes.
Lysis	Destruction of a bacterium as by bacteriophase, with the multiplication of phase particles in the process.
Lysozyme	Enzyme which hydrolyze the bond between N-acetylglucosamine and N-acetylmuramic acid of the peptidoglycan of the bacterial cell wall; this causes the cell wall to break and lysis of the cell usually follows.
Magnetic separator	The separation of seeds in this machine is done on the basis of magnetic properties. Seeds are selectively pretreated with magnetic material, *e.g.,* finely ground iron power.

Term	Definition
Maintainer line	The 'B' line
Male sterile	Producing no functional pollen.
Maleic hydrazide (MH)	1,2-dihydro-3, 6-pyridazinedione.
Malic acid fermentation	A metabolic process that results in the degradation of malic acid to ethanol.
Marine ecology	It deals with the marine habitat (ocean) and plants and animals and its relationship with environment.
Mass spectrometry	A process, which sorts sterams of electrically, charged particals according to their different masses by using deflecting fields.
Mechanical ventilation	Process of supplying or removing air with a mechanical device to or from any space, such as the interstitial space between products.
Medium	The supporting substance on or in which plants fungi and bacteria are cultured or grown. The term usually includes the nutrients as well.
Meiosis	A two stage type of cell division in sexually reproducing organisms that results in gametes with half the chromosome number of the original cell; or a special type of cell division found in gamete production. It consists of two divisions, one of which is reductional. Homologous chromosomes pair and assort at random to produce gametes with the haploid number chromosomes.
Melanin	A black or brown pigment of animal origin. In albinos a mutant recessive gene blocks the production of melanin.
Melanin	The pigment of the skin and hair. It is derived from tyrosine through the action of the enzyme tryosinase to produce DOPA (3,4-dihydroxy-phenylalanine).
Membrane	A selectively permeable boundary surrounding the cell or surrounding various sub-cellular particles.
Mendel's law	The principle that hereditary characters are determined by discrete particles (genes) that segregates at random in gamete formation.

Term	Definition
Mendelian population	A naturally breeding unit, isolated by some mechanism from other units of sexually reproducing plants or animals.
Mericarp	One of the two carpels of the fruit of an umbeliferous plant.
Mesocotyl	In some monocotyledons the part of the seedling stem which elongates below the shoot.
Mesosphere	It is third layer of atmosphere next to stratosphere.
Metabolism	The sum of all the chemical process occurring in a cell.
Metabolism	The sum of all the biochemical changes that take place in a living organism (anabolism and catabolism).
Metabolite	Any intermediate compound produced during the catabolism of a biological substance.
Metaphase	The stage in the cell division at which the chromosomes are shortened and arranged on an equatorial plate in the center of the cell.
Micelle	A very fine colloidal dispersion of polar lipids usually as a parallel array within a lipid phase, which fails to form a true solution of the molecules.
Michaelis constant	The constant, Km, derived from the Michaelis Menten equation, which expresses the substrate concentration necessary to produce one-half of the maximum velocity in an enzymatic reaction. $$Km = \frac{K_2 + K_3}{K_1}$$
Michaelis-Menten equation	$$Vo = \frac{V_{max}[S]}{Km + [S]}$$
Michaelis-Menten model	$$E + S \underset{K2}{\overset{K1}{\rightleftharpoons}} ES \xrightarrow{K3} E + P$$
Micron (μ)	A unit of length equal to 10^{-4} cm.
Micropyle	The opening from outside the seed leading through the integuments to the nucellus; it marks the position of the radicle.

Term	Definition
Microsomes	A subcellular particulate fraction usually containing ribosome and endoplasmic reticulum. Experimentally, it is the material sedimenting in an ultracentrifuge at a force of bout 100,000 x gravity.
Microspore	The "small spore" an asexual spore, the pollen grain in plants, in which develop the male gametes, the sperm.
Minimum lethal dose	The minimum dose of an agent sufficient to cause 100 per cent mortality of the test population.
Mist-O-matic treater	A type of machine used for treating seeds. It applies treatment as a mist directly to the seed.
Mitochondria	Membrane-surrounded structures, which contain the respiratory enzyme systems and electron transport chain. Mitochondria are found in the cytoplasm of cells.
Mitosis	A process of cell division in eukaryotic cells conventionally divided into the growth period (inter phase) and four stages: prophase, metaphase, anaphase and telophase. The stages conserve chromosome number by equally allocating replicated chromosome to each of the daughter cells.
Mitosis	The process whereby a cell nucleus divides into two daughter nuclei, each having the same genetic complement as the parent cell.
Moisture content	The moisture content of a sample is either the loss in weight when it is dried, or the quantity of water collected when it is distilled. It is expressed as a percentage of the weight of the original samples.
Molality	The expression of the number of moles of solute per litre of solution.
Molar (M)	A solution, which contains one mole of a compound to give one litre of final solution.
Monocotyledons	A group of plants so classified because the embryo usually has one cotyledon.

Term	Definition
Monoecious	Having male and female reproductive organs in the same individual.
	Having stamens and pistils in different flowers on the same plant.
Monosaccaride	Only one unit of sugar, *e.g.*, glucose, fructose, mannose.
Monosomic	Having a full set of chromosome minus one chromosome, 2n – 1.
Morphology	The study or science of structure at any level of organization (sub cellular, cellular, tissue, organ or gross structure of organisms).
Multiline variety	Multiline variety consists of two or more near isogenic lines of normally self-fertilizing plants. Growing the component lines separated and copositing them to constitute the breeders class of seed derive a mutiline.
Multiple alleles	More than the normal two alleles at a locus in the chromosome.
Mutable gene	An unstable gene that mutates frequently.
Mutant	A variant from the normal or wild type that is inherited in a Mendelian manner.
Mutation	Chemical modification of a gene such that one or more of the bases in the DNA is deleted or altered or a new base is inserted, thus changing the base sequence.
Mutator gene	A gene that causes other genes to mutate, *e.g.*, Dt. In corn causes a to change to A.
Muton	A subdivision of the gene, the smallest element, alteration of which can be effective in causing a mutation.
Myoglobin	A heme protein that is abundant in muscles.
Natural ventilation	Process of supplying or removing air using the natural forces of wind and temperature difference to or from any space such as interstitial space between products.
Nearest-neighbour frequency	In nucleic acid structure, the relative number of times each of the four nucleotides is found adjacent to any given nucleotide.

Term	Definition
Necrotic	Death of plant cells, especially when it results in the tissue becoming dark in colour.
Negative control of gene transcription	Negative control of transcription is the situation when a bound repressor protein prevents transcription of structural genes.
Nematodes	Threadlike round worms that live in soil and water and belong to the nemathelminthes.
Nick	The two parents for producing hybrid seed are said to nick when they produce high yields of seed of a highly productive and desirable hybrid.
Nick translation	A technique for radioactively labeling DNA molecules to high specific activity in order to produce a probe which may be used in some hybridization technique such as Southern blotting.
Nitrate nitrogen	Nitrate nitrogen is formed by the oxidation of nitrites and represents the most stable form of nitrogen. It is an indication of stability and is a determination of the completeness of the biological decomposition process.
Nitrite nitrogen	Nitrite nitrogen is formed by bacterial oxidation of ammonia nitrogen. It is not present in fresh wastes but appears after bacterial activity has taken place. The presence of nitrite nitrogen indicates that the waste has undergone partial decomposition and is unstasble. Nitrites can either be reduced back to ammonia or oxidized to nitrates.
Nitrogen	The unique element in amino acids, purines, pyrimidines, and other nitrogenous biomolecules, the major component of the atmosphere, and the element on which the human food supply is most dependent.
Nitrogen balance	The nutritional state of an individual in relation to his or her protein metabolism; it is a measure of the difference between ingested and excreted nitrogen.
Nitrogen fixation	The assimilation of atmospheric nitrogen by certain prokaryotic in to nitrogenous compounds that can be directly used by plants.

Term	Definition
Nobbe trier	It is a pointed tube, long, enough to reach the center of the bag, with an oval hole near the pointed end. It is made in different dimensions to suit various kinds of seeds.
Non-polar molecule	A molecule in which the electrons are evenly distributed over the structure in such a fashion that there is no separation of charge.
Normal seedlings	The seedlings, which show the capacity for, continued development into normal plants when grown in good quality soil and under favourable conditions of water supply, temperature and light.
Northern blotting	Northern blotting is analogous to southern blotting except that the sample nucleic acid that is separated by gel electrophoresis in RNA rather than DNA.
Notified kind/variety	In relation to any seed this means any kind or variety thereof notified under Section 5 of Seed Act, 1966.
Noxious weed seed	Seeds from any plant of those species considered being extremely destructive or harmful to agriculture. Law designates these species for Seed Law Enforcement.
Noxious weed seed (prohibited)	Seeds of noxious weed species restricted by law from being present in any quantity in agricultural seeds.
Noxious weed seed secondary	Seeds of noxious weed species permitted by law to be present in limited numbers in agricultural seeds.
Nucleic acid	A biological molecule (such as RNA or DNA) that allows organisms to reproduce; polymers compound of monomers called nucleotides joined by covalent bonds (phosphodiester linkages) between the phosphate of one nucleotide and the sugar of the next nucleotide.
Nucleoside	A hydrolytic product of nucleic acids containing a heterocyclic nitrogen base attached to a pentose, generally ribose or deoxyribose.
Nucleotide	A phosphate ester derivative of a nucleoside.

Term	Definition
Nucleus	The nucleus is a store of the cell's genetic information such as DNA in Chromosomes.
Nucleus seed	A group of progenies of individual plants taken at random froma variety for the purpose of purifying that variety of mixtures and off-types.
Nucleus Seed	It is original or first seed (= propagating material) of a variety available with the producing breeder (who developed in variety in question.)
Nullisonic	The complete lack of a given chromosome pair; $2n - 2$.
Nutrient	An element or compound that is needed by an organism to provide it with energy and a building material for growth.
Nutrients	Chemical substances in food that nourish the body by providing energy, building materials and factors to regulate needed chemical reactions in the body.
Nutrition	The council on food and nutrition of the American Medical Association defines nutrition as "the science of food; the nutrients and the substances therein; their action, interaction and balance in relation to health and diseases and the process by which the organism (*i.e.,* body) ingests, digest, absorbs, transports, utilizes and excretes food substances".
Objectionable weed	Weed plant whose seed is difficult to separate once mixed with crop seed and which is poisonous or injurious or has a smothering effect on the main crop. It is difficult to eradicate once established has a high multiplication ratio thus making its spread quick and serves as an alternate host for crop diseases and pests.
Off type	Plant or seed deviating significantly from the characteristics of a variety as described by the breeder in any observable rspect.
Official sample	It is submitted and drawn by quality control officer (QCO) or seed law enforced officer from selected containers of certified seeds/fertilizers, fungicides, insecticides, pesticides to confirm their quality as per labels or information given on containers.

Term	Definition
Okazaki fragments	DNA synthesis proceeds in a $5' \rightarrow 3'$ direction on each strand of the parental DNA. On the strand with $3' \rightarrow 5'$ orientation (the leading strand) the new DNA is synthesized continuously. On the strand that has $5' \rightarrow 3'$ orientation (the lagging strand) the DNA is synthesized discontinuously as series of short okazaki fragments that are then joined together. The small fragments are called Okazaki fragments. The new DNA strand, which is made by this discontinuous method, is called the lagging strand.
Oligonucleotide	A group of several nucleotides bound together in phosphodiester linkage between the phosphate of one nucleotide and a hyderoxyl group on the sugar of an adjacent nucleotide.
Omega-3 (ω-3) fatty acid	A fatty acid with itsfirst double bond first appearing at the thrid carbon atom from the methyl end ($-CH_3$).
Open pollination	Open-pollinated seed is seed produced a result of natural pollination as opposed to hybrid seed produced as a result of countrolled-pollination.
Operon	An operon is a coordinated unit of genetic expression.
Organic nitrogen	All nitrogen present in organic coumpounds is considered to be organic nitrogen. The nitrogen containing organic compounds are derivatives of ammonia, the oxidation of which forms ammonia nitrogen.
Origin (seed)	State or country where seed is grown.
Osmosis	It is essentially a special type of diffusion of liquids. When two solutions of different concentrations are separated by means of semi-permeable membrane, the diffusion of water or the solvent from the solution of lower concentration, until a state of dynamic equilibrium is attained, is known as osmosis.
Osmosis	Water molecules diffuse from region of their higher to region of their lower concentration through differentially permeable membrane.

Term	Definition
Osmotic pressure	Pressure required to stop osmosis when a solution is separated from pure water by a semipermeable membrane is termed the osmotic pressure.
Osteoblasts	Cells responsible for forming new bone tissue.
Osteoclasts	Cells responsible for bone destruction by resorbing the calcium from bone.
Other crop seed	Seeds of plants, which are grown as crops, other than main crop.
Other seeds	Other seed include seed and seed like structures of any plant species other than that of pure seed.
Out breeding	Mating of unrelated individuals or of individuals not closely related; the opposite of inbreeding.
Out cross	A cross usually natural to be a plant of a different genotype; mating of hybrid with a third parent; an off-type plant resulting from pollen of different sort contaminating a seed field.
Ovary	The female reproductive organ in which eggs are produced. In plants, the overy, containing ovules is at the base of the pistil.
Over dominance	An effect in the heterozygote, A/a greater than in the homozygous dominant, A/A.
Ovule	The megasporangium of a seed plant. After fertilization it becomes the seed.
	The body within the ovary of the flower that becomes the seed after fertilization and development.
	Ovule in the part of pistil/female organ in perfect/pistilate flower which contains an embryo with food reserve and protective coat. Mature intugmented megasporangium.
Oxidase	A class of enzyme that is active in oxidation/reduction reactions, which use oxygen as an electron acceptor.
Oxidation	A chemical change involving a loss of electrons.
	A type of chemical reaction involving the removal of electrons. Frequently, but by no means necessarily, the element oxygen is involved.

Term	Definition
Oxidative phosphorylation	Oxidative phosphorylation is ATP synthesis linked to the oxidation of NADH and $FADH_2$ by electron transport through the respiratory chain. Approximately three ATP molecules are synthesized per NADH oxidized and approximately two ATPs are synthesized per $FADH_2$ oxidized.
Oxidizing agent	In one sense, a substance capable of capturing an electron from another compound.
P/O ratio	In oxidative phosphralyation (acerobic respiration) the ratio of phosphate incorporated into ATP to the oxygen consumed.
Packaging	The process of filling, weighing and sewing bags.
Palea	The tiny bract with which the lemma encloses the flower in grasses.
Pallets	Wood frame structures, which are used to support a stack of bags, to prevent them from touching the floor and absorbing moisture.
Pantothenic acid	A vitamin that play important biochemical role as part of the structure of coenzyme A.
Paracentric inversion	A rotation of a segment of chromosome a full 180°, with the centromere beyond inversion, which is all within one arm of chromosome.
Parmutation	A mutation in which one allele in heterozygous condition changes permanently its partner allele.
Parthenogenesis	Reproduction without fertilization.
Parts per million (ppm)	Desingates the quantity of a substance contained in a million parts of a mixture or solution in a carrier such as air or water.
Passive transport	The movement of molecules across a membrane by passive transport does not require an input of metabolic energy. The molecule moves from a high concentration to a lower concentration, *e.g.* transportation of water, gases and urea.
Pathogen	Any organism capable of causing disease in a particular host or range of hosts. It obtains its nutrients wholly or in part from another living organism.

Term	*Definition*
PCR application	PCR has made a huge impact in molecular biology, with many applications in areas such as cloning, sequencing, the creation of specific mutations, medical diagnosis and forensic medicine.
Pectic material	Chemical substances derived from pectin, a jellylike substance found in fruits and other parts of plants; the chief cementing material of plant cells.
Pedicel	The stalk of a spikelet, flower stalk.
Pedigree	The genotype of an individual; a chart showing the ancestral history of an individual.
Pellagra	A niacin deficiency disease.
Percentage germination	The perventage germination on the seed anslysis certificate indicates the proportion of the number of seeds, which have produced seedlings, classified as normal under the conditions and within the period specified.
Perennial	A plant that produces vegetative growth year after year without the necessity of replanting.
Pericarp	The outer layers of cells surrounding the seed.
	The mature ovary wall.
pH	The negative logarithm of the hydrogen ion concentration.
	A numerical representation of the hydrogen ion concentration of a fluid that shows the acidity or alkalinity of the fluid.
pH index	The pH index is a number used to express the hydrogen ion concentration of a solution.
Phagocytosis	Phagocytosis is the uptake of large particles (bacteria and cell debris). The particle binds to receptors on the surface of the phagocytic cell and the plasma membrane then engulfs the particle and ingests it via the formation of large endocytic vesicle, a phagosome.

Term	Definition
Phenotype	The appearance of an individual produced by the genotype in co-operation with the environment.
	The observed character of an individual without reference to its genetic nature. Individuals of the same phenotype appear alike but may not breed alike.
Phenylketonuria	A disease inflicting serious brain damage in infants, caused by a recessive gene. It renders a child unable to metabolize phenylpyruvic acid, which accumulates in the brain.
Phosphoenolpyruvate (PEP)	A high-energy phosphate compounds that is an intermediate in both glycolysis and gluconeo-genesis. Phosphoenolpyruvate has a much larger standard free energy of hydrolysis than ATP.
Phosphorylation	The process whereby a phosphate group is attached to a substrate for example when fructose 1-phosphate becomes fructose 1-6, diphosphate during glycolysis.
Phosphorylation	The first step of glycolysis consists in the combination of hexose sugars with phosphates forming various types of hexose phosphates and the process is called phosphorylation of hexose molecules.
Photo-phosphorylation	The process of generating ATP from ADP and phosphate by means of a proton-motive force generated by the thylakoid membrane of the chloroplast during the light reactions of photosynthesis.
Photo-respiration	A metabolic pathway that consumes oxygen, evolves carbon dioxide, generates no ATP, and decreases photosynthetic out put; generally occurs on hot, dry, bright days, when stomata close and the oxygen concentration in the leaf exceeds that close of carbon dioxide.
Photochemical	Part of the radiant energy absorbed by chlorophyll is used in producing a chemical change, that is the effect is photochemical.
	Pertaining to a chemical reaction activated by light.

Term	Definition
Photoperiod	The number of hours of daylight needed by a plant before it will begin to flower.
Photoperiodism	The action of certain amounts of light stimulating floral initiation in plants.
Photophsophorylation	The interaction of the weak oxidant of PS I and weak reductant of PS II results in the production of ATP. This process is therefore, also known as photosynthetic phosphorylation or photophosphorylation.
Photoreaction	A reaction that is initiated or hastened by light.
Photosynthesis	The conversion of light energy into chemical energy and the use of that energy to form carbohydrate and other cellular constituents from carbon dioxide and water.OR Photosynthesis is a chemical process it uses solar energy to synthesize carbohydrate from carbon dioxide and water. In the light reactions, the light energy derives the synthesis of NADPH and ATP. In the dark reactions (carbon-fixation reactions), the NADPH and ATP are used to synthesize carbohydrate from CO_2 and H_2O.
Photosystem I	A photochemical reaction system in photosynthesis; coupled with photosystem II.
Photothermal	Pertaining to combine effects of light and temperature.
Phototropism	Growth of a plant shoot toward or away from light.
Photsystem II	A photochemical reaction system in photosynthesis; coupled with photosystem I.
Physical dormancy	Due to presence of growth inhibitors or absence of growth promoters. Presences of ABA and Auxins in low concentration inhibit the germination.
Physiological process	Molecules are transported across the membrane against concentration gradient at the expences of metabolic energy (ATP).
Physiology	The study of the functions and activities of seed.

Term	Definition
Phytotoxic	Poisonous to plant.
Picking tables	These machines are used for making hand separations that cannot be made mechanically. This machine consists of a feed hopper with an adjustable gate and a horizontal endless belt.
Pigment	Substance that appears coloured by virtue of differntial absorption of radiant energy. These substances impart colour to tissue of plants. Green colour is a result of chlorophyll; orange and some red colours are due to many carotenoids; many red to blue colours are anthocyanins, light yellow colours are flavones.
Pinocytosis	Pinocytosis is the nonspecific uptake of extracellular fluid via small endoplastic vesicles that pinch off from the plasma membrane. This is a constitutive process occurring in all eukaryotic cells.
Pinocytosis	Cell membrane takes in fluid droplets by forming vesicles around them.
Pistil	The ovule-bearing organ of a flower at the base of the style.
	The seed-bearing organ of flower.
Pistillate	Bearing female gametes only have a female type with no stamens.
Plant Biochemistry	It deals with chemical manifestation of plant metabolism. Plant biochemist interested in studying properties of plant kingdom, *e.g.,* Photosynthesis, chlorophyll synthesis, energy biosynthesis and chemical compounds biosynthesis.
Plant breeder	Person or organization actively engaged in the breeding and maintenance of varieties of plants.
Planting ratio	The recommended ratio in which the male and female parental lines are planted to make a crossing block in hybrid seed production.
Planting stakes	Stem cuttings used for planting.
Plasmagene	A unit in the cytoplasm causing hereditary traits, *e.g.,* kappa in paramecium.

Term	Definition
Plasmid	A small ring of DNA that carries accessory genes separate from those of a bacterial chromosome.
Plastids	Small bodies of specialized protoplasm, especially in plant cells, *e.g.*, chloroplasts containing the green pigment chlorophyll.
Plenum	An air chamber from which air is distributed or collected. It is frequently a main duct of a drying system, where the velocity pressure is reduced to a desired minimum.
Plumule	The major young bud of the embryo within a seed or seedling from which will develop the aerial protions of the plant. It usually occurs at the tip of stem-like structure called the epicotyl; that part of the embryonic plant axis above the cotyledons.
Pneumatic separator	A see processing machine in which seed separations are made by use of air on the basis of differences in terminal velocity. It has a fan on the intake end and a pressure within the machine, greater than normal atmospheric pressure is buit up. The air blast makes the seed separation.
Polar	A substance having one region more negatively charged than another region.
Polishers	The machine used for polishing seeds to improve their luster.
Pollen grain	The microspore in plants containing a tube nucleus and a generative nucleus that divides either in pollen grain or in pollen tube to form two sperm nuclei.
Pollen parent	The parent that furnishes the pollen, which fertilizes the ovules of the other parent in the production of seed.
Pollen parent (male parent)	The plant supplying the pollen for a hybrid.
Pollen shedder	In hybrid seed production involving male sterility the plants of 'B' line present in 'A' line are tremed as pollen shedders.
	In hybrid seed production programme, the plant of 'B' line/or plant which having viable pollens on anthesis present in 'A' (CMS line) is termed as pollen shedder.

Term	Definition
Pollination	The transference of pollen from the anthers to the stigmas of flowers.
Polygene	One of many genes necessary for a given phenotypic effect as found in quantitative inheritance.
Polymers	Polymers are composed of many copies of a few small molecules linked in chains by covalent bonds. These subunits of polymers are referred to as monomers or residues.
Polypeptides	Polypeptides are polymers composed of amino acids connected by peptide bonds. It can be described in four groups: 1. Primary structure, 2. Secondary structure, 3. Tertiary structure, and 4. Quaternary structure.
Polyploid	Composed of more than two genomes or chromosome sets. 2n = diploid, 3n = triploid, 4n = tetraploid etc.
Polysaccharide	A polymeric sugar yielding many monosaccahride units upon hydrolysis; or at least 15 different sugars are used to form various polysaccharides. The monomeric sugars can be bonded to one another in multiple ways; thus, many polysaccharides are nonlinear; branched molecules.
Population	A group of individuals of the same species is known as population.
Population genetics	The branch of genetics concerned with the frequencies of genes (alleles) in a population.
Positive control of gene transcription	Positive control of gene transcription is when the regulatory protein (an activator) binds to DNA and turns on transcription.
Potential energy	The energy that usually concerns us when we study biological or chemical systems is potential energy or stored energy.
Precision grader	This machine is used to upgrade seed quality. It makes separations on the basis of differences in width and thickness.

Term	Definition
Pre-emergence	Before emergence often refers to the treatment of the soil with weed control chemicals after palnting and before the crop plants appear above ground.
Pressure system	Method of air movement for drying in which the air forced through the product by the air duct or ducts being at a pressure above atmospheric pressure. Forcing system of air movement.
Pre-treatment	Any physical or chemical laboratory treatment of the working sample that is given solely to facilitate testing.
Previous crop	The crop grown in the season immediately preceding the one in which the seed crop is grown.
Primary dormancy	Due to inherent properties of seed or due to unfavourable environmental conditions – 2 types Exogenous/enforced/relative dormancy: due to absence of essential germination components like water, light, temperature etc. Endogenous/innate/true dormancy: due to inherent properties of seed.
Primary leaf	The just true leaf of the seedling.
Primary root	The root developing directly from the radicle.
Primary sample	A primary sample is a small portion taken from one point in the lot.
Primary structure of protein	The linear sequence of amino acids joined together by peptide bonds is termed the primary structure of the protein.
Principles of PCR	The polymerase chain reaction allows an extremely large number of copies to be synthesized of any given DNA sequence provided that two oligonucleotide primers are available that hybridize to the flanking sequences on the complementary DNA strands. The reaction requires the target DNA, the two primers, all four deoxyribonucleotide triphosphates and a thermostable DNA polymerase such as Taq DNA polymerase. A PCR cycle consists of three steps, denaturation, primer annealing and elongation. This cycle is repeated for a set number of times depending on the degree of amplification required.

Term	Definition
Probability	The chance or likelihood of a given possible event.
Processing	It refers to all steps involved in the preparation of harvested seed for marketing namely handling, shelling, pre-conditioning, drying, cleaning, size grading, upgrading, treating and packaging. In a narrow sense, however, it refers only to pre-conditioning, cleaning, size grading and upgrading of seeds.
Progeny	Offspring, plants grown from the seeds produced by parent plant.
Prokaryotens	Simple cells having only a singe membrane.
Prokaryotes	Prokaryotes (bacteria and blue-green algae) are the most abundant organisms on earth. A prokaryotic cell does not contain a membrane-bound nucleus.
Prophase	A stage of cell division in which chromosomes are first visible as chromosomes.
Prosthetic group	A non-protein component attached to a protein that is generally necessary for the proteins biological activity.
Protamines	In sperm heads, DNA is particularly highly condensed and here the histones are replaced with small basic proteins called protamines.
Protein	A three dimensional biological polymer constructed from a set of 20 different monomers called amino acids; or proteins are synthesized in the amino-to-carboxyl direction by the sequential addition of amino acids to the carboxyl end of the growing peptide chain.
Protein efficiency ratio (PER)	A term used expressing the growth promoting ability of a protein, *i.e.,* gain in weight per gram of protein consumed.
Proton pump	An active transport mechanism in cell membranes that consumes ATP to force hydrogen ions out of a cell and in the process generates a membrane potential.
Prototroph	A wild type bacterium that will grow on a minimal medium.

Term	Definition
Provenance	The region where seed was harvested.
Provisional certificate	A certificate issued before the completion of a test or tests and which includes a statement that a final certificate will be issued on completion.
Psychrometric chart	The device, which simplifies the measurement of air properties and eliminates many time-consuming and tedious calculations, which would otherwise be necessary.
Pubescence	A hairy covering usually of short soft hairs.
Pure line	A strain in which all individuals have descended by self-fertilisation from a single homogygous individual. A pure line is genetically pure.
Pure liveseed	Percentage of pure germinating seed determined by multiplying by the pure seed percentage by its own germination percentage and dividing the product by one hundred.
Pure seed	The pure refers to the seeds of the species stated by the sender or found to predominate in the purity test. It includes all botanical varieties and cultivars of that species even if immature, undersized, shriveled, diseased or germinated, provided they can be definitely identified as of that species.
Purine	One of two families of nitrogenous bases found in nucleotides consisting of two members: adenine (A) and guanine (G).
Purity	The composition by weight of the sample being tested and by inference the composition of seed lot; the identity of various species of seeds and inert particles constituting the samples.
Pyridine nucleotide	Both nicotinamide adenine dinucleotide (NAD^+) and nicotiamine adenine dinucleotide phosphate ($NADP^+$).
Pyrimidine	One of two families of nitrogenous bases found in nucleotides consisting of three members: cytosine (C), thymine (T) and uracil (U).
Pyruvate	A key intermediate in the extraction of energy from glucose, some amino acids and glycerol.

Term	Definition
Quantum	Energy of an electromagnetic quantum calculated from its wavelength expressed in millimicrons, m: $eV=1239.8/m$.
Quaternary structure of protein	If a protein is made up of more than one polypeptide chain it is said to have quaternary structure. This refers to the spatial arrangement of the polypeptide subunits and the nature of interactions between them.
Radicle	The rudimentary root of the embryo. It forms the primary root of young seedling.
Radioactive	Having a nucleus that gives off particles and energy; characteristic of unstable isotopes of a chemical element.
Rate constant	Proportionality constant between the rates of a reaction on the concentration of species, which influence the rate.
Rate law	An equation expressing the dependence of the rate of a reaction on the concentration of species, which influence the rate.
Recessive	A term coined by Mendel to describe characters, which recede completely in the F_1. Action of the recessive allele is suppressed by the dominant.
Reciprocal crosses	Hybrids made by using the mutant type as female in one case and as male in the other.
Recombination	A new combination of linked genes other than the parental types, *e.g.*, in an F_1 of AB/ab genotype. Both Ab and aB gametes would represent recombination.
Red drop	Quantum yield of photosynthesis decreased under greater wavelength than 680 nm in the red zone is known as red drop.
Redox potential	The oxidation-reduction potential, E (or redox potential) of a substance is a measure of its affinity for electrons. The standard redox potential (E'o) is measured under standard conditions, at pH 7.0 and is expressed in volts.
Reduction	The gaining of electrons by a substance involved in a redox reaction.

Term	Definition
Registered seed	The progeny of breeder's select or foundation seed handled under procedures acceptable to the certitying agency to maintain satisfactory at that temperature.
Relative humidity	The amount of water present in the air at a given temperature in proportion to its maximum water holding capacity at that temperature.
Release	Refers to the release of varities by appropriate authorities such as the Central Variety Committee; State Variety Release Committee.
Repressible enzymes	Enzymes whose rate of production is decreased when the intracellular concentration of certain metabolites increases.
Resistance to air flow	The property of an air passage between the grains makes it necessary to apply pressure to get airflow. When used quantitatively, the dimensions as used should be made clear to the user. Resistance to air flow is usually expressed in term of inches of water per foot of depth at a certain airflow for a particular product.
Respiration	The oxidative breakdown and releases of energy from fuel molecules by reaction with oxygen in aerobic cells.
Respiratory control	The control of respiration by limiting levels of the acceptor molecule ADP.
Respiratory quotient (RQ)	Moles of CO_2 produced ÷ Moles of O_2 consumed
Restorer line	An inbred line that, when crossed on a male sterile strain, causes the resulting hybrid to be male fertile (R line).
Restriction enzyme	Restriction enzymes allow DNA to be cut at specific sites, nucleic acid hybridization allows the detection of specific nucleic acid sequences, DNA sequencing can be used to easily determine the nucleotide sequence of a DNA molecule; or A degradative enzyme that recognizes and cuts up DNA (including that of certain phases) that is foreign to a cell.

Term	Definition
Restriction fragment length polymorphisms (RFLPs)	Differences in DNA sequence on homologous chromosomes that result in different patterns of restriction fragment lengths (DNA segments resulting from treatment with restriction enzymes); useful as genetic markers for making linkage maps.
Reverse mutations	Mutations from a mutant type to the normal or wild type allele. Less frequent than the mutation from the wild type to the mutant.
Reversible process	A process occurring so slowly in a system that the system is at all times in a state of equilibrium. The process can be stopped and reversed just by making an infinitesimal change in the external conditions.
Rhizome	A horizontal underground stem.
Ribosome	A cytoplasmic nucleoprotein particle, consisting of RNA and protein. Ribosomes are the site of protein synthesis.
	A highly organized assembly of RNA and protein molecules present in the cytosol.
RNA (ribonucleic acid)	Single strand molecules that are complementary copies of a segment of DNA. Some RNAs are in the information carrying business and some are in the protein building business.
RNA synthesis	RNA synthesis differs from that of DNA in several ways. RNA polymerase does not require a primer. The DNA template is fully conserved in RNA synthesis. Whereas it is semi conserved in DNA synthesis. RNA polymerase has no known nuclease activities. The growth of an RNA chain is in the $5' \rightarrow 3'$ direction as in DNA synthesis. RNA polymerase moves along the DNA template strand in the $3' \rightarrow 5'$ direction, since the DNA template strand is anti-parallel to the newly synthesized RNA strand. The sequence of long RNA chains can be elucidated by the use of specific hydrolytic enzymes and finger printing methods.

Term	Definition
Rock-it corn grader	A width and thickness grader which uses flat screens to make width and thickness separations.
Rodewald apparatus	This consists of a zinc box covered with glass in which seeds are exposed to direct or diffuse light. The bottom of the apparatus contains water, over which a layer of moist sand is placed on a tray. The seedbed consists of unglazed porcelain dishes placed in the moist sand or the dishes may be placed directly in the water. The temperature of water is controlled thermostatically by electricity.
Rogue	An off-type plant; undesirable plant.
Roguing	The removal of individual plants, which deviate in a significant manner from the normal or average type of a variety. A step in the maintenance of purity in an established variety or in the attainment of purity in a new variety; the act of removing nderirable plants.
Role of citric acid cycle	The main role of the citric acid cycle is the oxidation of pyruvate (formed during the glycolytic breakdown of glucose) to CO_2 and H_2O with the concomitant production of energy. It also has a role in producing precursors for biosynthetic pathways.
Role of fatty acids	They are components of membranes (glycerophospholipids and sphingolipids). Several proteins are covalently modified by fatty acids. They act as energy stores (triacylglycerols) and fuel molecules. Fatty acid derivatives serve as hormones and intracellular second messengers.
Room germinator	It is a modification of the cabinet. It is constructed on the same principles but is large enough to permit workers to enter it and place the tests along either side of a central passage.
Root hairs	A fine tubular outgrowth of a surface cell of the root.
Root slips	The plants arising from portion of a root, when bedded for propogation.
Rouging	Removal of off type plants, plant showing variation in expression, abnormal plants, pollen shedder from the seed production plot.

Term	Definition
Rules (seed testing)	The ISTA rules for testing seeds.
Sampling	The process of obtaining a sample of a size suitable for tests, in which the same constituents are present as in the seed lot and in the same proportions. The sample is obtained from the seed lot by taking small portions at random from different positions in the lot and combining them. From this sample smaller samples are obtained by one or more stages.
Sampling intensity	Refers to the number of primary samples taken from the seed lot.
Saturated fatty acids	Fatty acids that contain only single carbon-carbon bonds.
Scalper	A seed-processing machine designed to remove the bulk of trash from seed so as to facilitate elevating and processing operations. It is basically a vibrating or rotating screen or sieve.
Scarifier huller	The machine used for scarification of hard seeds.
Scarify, scarification	The process of mechanically scarring or roughing the hard seed coat to make it more permeable to water.
Schizocarp	A fruit derived from a compound pistil in which the one seeded carpels separate from one another at maturity as in umbelliferae.
Sclerotia	A firm frequently rounded mass of hyphae with or without the addition of host tissue and normally having spores in or on it.
Screen dams	Strips of wood, fastened across the screen to retard flow of seed down the screen.
Screening DNA libraries	Genomic and cDNA libraries can be screened by hybridization using a labeled DNA probe complementary to part of the desired gene. The probe may be an isolated DNA fragment (*e.g.*, restriction fragment) or a synthetic oligonucleotide designed to encode part of the gene as deduced from a knowledge of the amino acid sequence of part of the encoded protein. In addition, expression cDNA libraries may be screened using a labeled antibody to the protein encoded by the desired gene or by using any other ligand that binds to that protein.

Term	Definition
Scurvy	A disease resulting from a deficiency of vitamin C.
Scutellum	The cotyledon of an embryo of a grass, a food absorbing structure.
Second law of thermodynamics	The principle where by every energy transfer or transformation increases the entropy of the universe. Ordered forms of energy are at least partly converted to heat and in spontaneous reactions, the free energy of the system also decreases.
Secondary dormancy	Induced by manipulation of physical and environmental factors.
Secondary structure of protein	It refers to the regular folding of regions of the polypeptide chain (α-helix and β-plated sheet).
Seed	The mature ovule containing a dormant plant embryo.
	A mature ovule consisting of an embryonic plant together with a store of food, all surrounded by a protective coat. In seed production, seed stands for any of the following classes of seeds used for swing or planting. 1. Seeds of food crops including edible oilseeds and seeds of fruit and vegetables. 2. Cotton seeds. 3. Seeds of cattle rodden. 4. Jute seeds and include, seedlings, tubers, bulbs, rhizomes, roots, cuttings all types of grafts and other vegetatively propogated material of food crops or cattle fodder.
	Seed is defined as matured (after fertilization) and ripened ovule which contains an embryo with food reserve and protective coat.
Seed analyst	Appointed under Section 13 of Seeds Act, 1966; the person supervising or carrying out the seed analysis work.
Seed blower	This equipment is used to separate lightweight material such as chaff, and empty florets in grasses, from the heavier seeds.
Seed borne	Carried on seeds.

Term	Definition
Seed coat	The outermost covering of a seed.
Seed disinfection	Riddding the seed surfaces of organisms which are potentially disease producing.
Seed disinfestations	Dipping seeds in chemicals, soaps, and fungisides to kill pathogens.
Seed health	Health of seed refers primarily to the presence or absence of disease causing organisms such as fungi, bacteria and viruses; and animal pests such as eelworms and insects; physiological conditions such as trace element deficiency may also be involved.
Seed inspector	Appointed under Section 13, of the Seeds Act. Their main function is to take samples from sale points where seeds of notified varieties are being sold, get them anaysed, if necessary and initiate prosecution against offenders of the Seeds Act.
Seed lot	A uniformly belended quantity of seed designated by proper unber or rank.
	A specified, uniformly blended quantity of seed which is physically identifiable for issuing certificate of quality seed is known as 'seed lot'.
Seed lot certificate (orange or green certificate)	The form of international Seed Anlysis Certificate used when the sample is drawn officially from the lot, under the authority of a member station, and the procedure followed identies the certificate with the seed lot.
Seed parent	The strainfrom, which seed, is harvested in the hybrid seed field. Also commonly used to designate the female parent in any cross-fertilisation.
Seed pellets	More or less spherical units developed for precision sowing, usually incorporating a single seed with the size and shape no longer readily evident. The pellet in addition to the pelleting material may contain pesticides, dyes or other additives.
Seed piece	This term is applied to pieces cut from stem tissue for purpose of vegetative multiplication.

Term	Definition
Seed protection	Treating the seed with chemicals, which by their presence prevent attack by seed-borne or soil-borne organisms on seed after planting.
Seed quality	Seed quality is a relative term and means the degree of excellence when compared to an acceptable standard. The seeds meeting required standards of purity, germination and other attributes are referred to as quality seeds.
Seed sample certificate (blue certificate)	The form of International Seed Analysis Certificate used when sampling from the lot is not under the responsibility of a member station (seed laboratory).
Seed stalk	The erect stalk on a plant that produces flowers and seeds. Applied particularly to root crops and leafy vegetable crops that produce seed after the desired product has fully developed.
Seed tapes	Narrow bands of material, such as, paper or other degradable material, with seeds placed in rows, groups or at random throughout the sheets.
Seed technology	The science which deals with production, harvesting, processing, testing, packaging, sorting and marketing of seeds. Seed technology is applied science deals with production of seed/plant parts by adopting proper technique, seed certification and seed test used for raising or propagation or new commercial crop.
Seed testing	The science of evaluating the planting quality of seed before it is sown.
Seed treatment	Application of fungicide/insecticides/both to seed so as to disinfect and disinfest them or to protect them from some pathogens and store grain pest is termed as seed treatment.
Seed viability	The capacity of seed or any plant part, (*e.g.* cuttings) to show living properties like germination and growth, i.e., normal seedling under favourable environmental conditions (preferably in absence of dormancy).

Term	Definition
Seed vigour	Seeds having better germination in laboratory and emergence in field, i.e., sum total of all properties of seed which determines the potential of activity and performance of seed or seed lot during seed germination and seedling emergence in field condition is considered as 'seed vigour'.
Seedling	A young plant developing from the embryo in seed.
	The embryo or young plant, from the time it emerge from the seed until it is entirely dependent on food manufactured by itself. It consists of an epicotyl, one or two cotyedons, hypocity and root. The single cotyledon is usually held within the seed coat in monocots.
Seeds Act	As used in this book, Seeds Act refers to the Indian Seeds Act, 1966; an Act to regulate the quality of seed offered for sale.
Select seed	Select seed is unique to the Canadian Certification system. It is the approved progeny of breeder's or select seed, produced in a manner to ensure its specific genetic identity and purity by those growers authorized by the certifying agency for the production of this class. Select seed is not a seed of commerce.
Selfed	Said of a pistil that is fertilized with pollen from the same plant that bears the pistil; also applied to seed resulting from such fertilization.
Self-fertilise	To fertilise the ovule of a flower with the pollen of the same flower.
Self-fertilization	Fertilization following application of plants own pollen.
Self-inocompatibility	Inability to set seed from application of pollen produced on the same plant.
Self-pollination	Pollinating a plant with its own pollen; selfing.
Self-sterility	In capability of producing seed when self-pollinated. Several alleles, S_1, S_2, S_3 etc. are responsible for this phenomenon.

Term	Definition
Self-tolerance	Cells that produce antibody that reacts with normal body components are killed early in fetal life so that the adult animal normally is unable to make antibodies against self, a condition called self-tolerance.
Seminal	Pertaining to seed or germ. Seminal organs are those already developed in the embryo within the seed.
Seminal roots	The primary root and a number of secondary roots arising from the embryo axis and forming the seedling root system in cereals.
Senescence	The terminal phase in the life cycle of an organism during which degradation gradually takes place.
Sephadex	Sephadex is one of the dextran gels used in affinity chromatography.
Serum	The liquid portion of coagulated blood from which all cells and fibrinogen have been removed.
Service sample	A sample submitted to the Central Seed Laboratory, or a State Seed Laboratory for testing, the results of which to be used as information for seedling, selling or labeling purposes.
Shedding tassel	Refers to the tassels in female parent rows shedding pollen or plant have shed pollen in hybrid maize plots. During field inspection, a tassel whose main spike or any side branch, or both have shed pollen or are shedding pollen, in more than 5 cm of branch length is counted as a shedding tassel.
Shikimic acid ($C_7H_{10}O_5$)	A crystalline acid that is a plant constituent and an intermediate in the biochemical pathway from phosphoenolpyruvic acid to tyrosine.
Shoot apex	Terminal portion of the shoot, that contains the main growing point.
Sib	Progeny of same parents derived from different gametes.
Sibbed	Mated individuals having the same parentage.

Term	Definition
Sickle cell anemia	A condition produced by an abnormal hemoglobin molecule in the homozygous condition.
	An inherited disease caused by an alteration in the beta chains of hemoglobin that results in a decreased ability of red blood cells to carry oxygen.
Sickle cell trait	Trait shown by individuals characterized as heterozygote for the sickle cell gene.
Silk (maize)	The stigma and style of the female maize flower and through which the pollen tube grows to reach the embryo sac.
Silo	Storage structures whose height is more than the width or diameter.
Single cross	The hybrid of two pure lines.
	A single cross is the first generation hybrid between teo specified inbred lines.
Single cross parent	The F_1 offspring of two inbred parents, which in turn is used as a parent, usually with another line or single cross parent to produce commercial hybrid seed, as in maize.
Single psychrometer	A single psychrometer has two thermometers mounted on a base plate. The one with the sock is the wet bulb thermometer; the other is the dry bulb. The wet bulb extends below the dry bulb. This is done purposely, so that the sock can be dipped in water without wetting the dry bulb thermometer.
Sleeve trier	A hallow brass tube inside a closely fitting outer shell or sleeve which has a soild pointed end. The tube and sleeve have open slots in their walls, so that when the tube is turned until the slots in the tube and sleeve are in line, seeds can flow into the cavity of the tube and when the tube is given a half turn the opening are closed. The tubes vary in length and diameter, being designed for different kind of seeds and various sizes of containers and are made with or without partitions.

Term	Definition
Slurry	Suspensionof wettable powder of fungicide and/or inecticide in water.
Slurry treater	The machine used for slurry seed treatment.
Sodium pump	An active transport mechanism associated with cell membranes. The pump moves sodium ions out of the cell aginst a concentration gradient producing the resulting potential across the membrane.
Sodium pump	The energy requiring process by which a high concentration of (Na^+-K^+ pump) potassium ions and low concentration of sodium ions are maintained inside nearly all animal cells.
Solution or crysatlloid	A solution or crystalloid is a mixture of two or more substances in which all the molecules or ions of one substance are separated from one another and dispersed throughout the medium of the other. The substance dispersed is known as solute and the substance in which the solute particals are dispersed is known as solvent.
Somatic cell	A cell in the body of the organism, with 2n chromosomes. The contrasting type is the germ cell, within chromosomes.
Somatic mutation	A mutation in a somatic cell.
Southern blotting	Southern blotting involves electrophoresis of DNA molecules in an agarose gel and then blotting the separated DNA bands on to a nitrocellulose filter. The filter is then incubated with a labeled DNA probe to detect those separated DNA bands that contain sequences complementary to the probe.
Species	A group of closely related organisms.
Spectrophotometer	An analytical device consisting of a light source, diffraction grating, sample holder and detection system that is used to measure absorption spectra or changes at fixed wavelength.
Spermatogenesis	Formation of mature sperm in animals.
Spiral separator	A machine used to upgrade seed quality. It makes separation on the basis of differences in relative ability to roll.

Term	Definition
Sporogenesis	The formation of microspores (pollen) and megaspores (embryo sac) in plants.
Sport	A mutation.
Stacking	Refers to the arrangement of seed bags in a manner such as to prevent their falling.
Stages of cell division	1. Interphase, 2. Prophase, 3. Metaphase, 4. Anaphase and 5. Telophase.
Stalk	A stem like supporting structure.
Stamen	The part of the flower bearing the pollen; pollenbearing organ; each stamen is composed of a stalk.
Staminate	Producing stamens only of a male plant.
Standard free energy change	The free energy change of a chemical process by which reactants at unit activity are converted into products at unit activity.
Starch	Starch is a mixture of unbranched amylose (glucose residues joined by $\alpha, 1 \rightarrow 4$ bonds) and branched amylopectin (glucose residues joined $\alpha, 1 \rightarrow 4$ with some $\alpha, 1 \rightarrow 6$ branch points). Starch is produced in the stroma of chloroplasts and stored there as starch grains. Starch synthesis occurs from ADP-glucose, CDP-glucose or GDP-glucose (but not UDP-glucose).
State functions	Properties of the state of a system, which developed only upon the condition of a system and not its previous history.
State seed laboratory	In any State, the State Seed Laboratory established or declared as such under sub-section (2) of Section 4 of Seeds Act, 1966.
Steady state	For an intermediate in a chemical reaction, a state in which the rate of breakdown of the intermediate is equal to the rate of its formation, with the result that its concentration does not change.
Steckling	Small sugar beet stored over winter and planted for the production of seeds.

Term	Definition
Stereoisomers	Substances which have the same molecular formula and order of attachment of atoms in the molecule but which differ in the three-dimensional geometry or configuration of the atoms.
	The D and L sterioisomers of sugars refers to the configuration of the asymmetric carbon atom furthest from the aldehyde or ketone group. The sugar is said to be a D isomer if the configuration of the atoms bonded to this carbon atom is the same as for the asymmetric carbon in D-glyceraldehydes.
Sterile	A plant that fails to set seeds even though compatible pollen is applied to the stigma of the flower. Cross-sterile plant fail to set seed with pollen from other plants, self-sterile plants produce no seed from their own pollen.
Sterility	Inability to produce offspring.
Stick or sleeve-type trier	A hallow brass tube inside a closely fitting outer shell or sleeve which has a soild pointed end. The tube and sleeve have open slots in their walls, so that when the tube is turned until the slots in the tube and sleeve are in line, seeds can flow into the cavity of the tube and when the tube is given a half turn the opening are closed. The tubes vary in length and diameter, being designed for different kind of seeds and various sizes of containers and are made with or without partitions.
Stigma	The upper end of the pistil that receives the pollen.
Stomatal transpiration	Most of the foliar transpiration takes place through the stomatal opening and is therefore, known as stomatal transpiration.
Stoner	The stoner is essentially a modified gravity separator designated to make a two-part separation by differences in specific gravity.
Strain	A type within a variety that constantly differ in genetic factors from other stains of the same variety. May become a variety.

Term	Definition
Stratosphere	The second layer of air mass extending about 30 km above tropopause is called stratosphere.
Strophiole	A swollen appendage at the hilum of some seeds.
Style	The stalk of the pistil between stigma and ovary
Submitted sample	The sample submitted to a seed-testing laboratory. It comprises the composite sample reduced as necessary.
	Required quantity of seed sample taken from composite sample for different seed tests is known as submitted sample.
Substrate	The substrate whose reaction is catalyzed by an enzyme; or the chemicals that undergo a change in a reaction catalyzed by an enzyme are called substrate of that enzyme.
Substrate	A substance that is acted upon as by an enzyme. Also a culture media.
Subtratum	The material upon which seeds are placed for a germination test.
Sucker	An offshoot that develops from an adventitious bud located on the roots or lower stem of a plant.
Sugar derivatives	Other groups to form a wide range of biologically important molecules including phosphorylated sugars, amino sugars and nucleotides can replace the hydroxyl groups of sugars.
Superoxide dismutase	An enzyme that can neutralize a superoxide free radical.
Supplemental heat	Adding a small amount of heat for a limited temperature rise, usually less than 20°F to dry within the maximum permissible time before spoilage.
Suppressor	One, which suppresses the action of another gene or other genes.
Surge bin	The bins mounted above the processing machine.
Suspension	When the particles of a matter do not separate into molecules and simply dispersed as such throughout the liquid, the system is called a suspension.

Term	Definition
Swollen seeds	Seeds which have imbibed water and although healthy in appearance have not germinated during the prescribed test period.
Symbiosis	An ecological relationship between organisms of two different species that live together in direct contact.
Synapses	The conjugation or pairing of homologous chromosomes at meiosis.
Synchronization	Simultaneous flowering of male and female parent in hybrid seed production. It is essential for better seed set.
Syndrome	A group of symptoms that occur together and characterize a disease.
Synecology	Synecology deals with system of many species-whole communities or major fraction of communities and ecosystems and its environment.
Synthesis of palmitate	It requires the input of 8 molecules of acetyl CoA, 14 NADPH and 7 ATP.
Synthetic	Refers to varieties produced by the combination of selected lines or plants and subsequent normal pollination.
Tailings	Partly threshed material that has passed through the coarse shakers or straw walkers of a threshing machine and has passed over the fine sieve.
Tassel	The flower cluster at the tip of a corn plant comparising pollen bearing flowers. The staminate inforscence of maize.
TCA cycle	A series of catabolic reactions taking place in the mitochondrial matrix.
Telophase	The last stage in cell division, in which the chromosomes are assembled at each end of the cell.
Temperature alternating	In germination testing, specific temperatures with seeds being held at both the lower and higher temperatures for a designated time each day.
Temperature constant	In germinationtesting a specific temperature which should not vary by more than 1°C.

Term	Definition
Tempering	Equalisation of moisture or temperature through-out kernel or products; to bring grain to a desired moisture or temperature for processing.
Template	A macromolecular mold for the synthesis of another macromolecule.
Temporary wilting or transient wilting	The leaves droop down during day time and regain their turgidity during night such situation is known as temporary wilting or transient wilting
Terminal bud	The shoot apex enveloped by several more or less differential leaves.
Terminalization	The movement of a chiasma away from the Centro mere and towards the end of a tetrad.
Tertiary structure of protein	Tertiary structure in a protein refers to the three-dimensional arrangement of all the amino acids in the polypeptide chain.
Testa	The hard outer covering of seed; seed coat.
Testcross	The cross of an F_1 by the homozygous recessive, useful in linkage studies. Contrast with backcross.
Tetrad	The group of four chromatids the results from pairing of homologous chromosomes and division of each chromosome into two chromatids.
Tetraploid	Having four genomes (4n) instead of two as in a diploid.
Tetraploid	A plant with four sets of identical or similar chromosomes.
The basis of DNA cloning	To clone into a plasmid vector, both the plasmid and the foreign DNA are cut with the same restriction enzyme and mixed together. The cohesive ends of each DNA reanneal and are legated together. The resulting recombinant DNA molecules are introduced into bacterial host cells. If the vector contains an antibiotic resistance gene(s) and the host cells are sensitive to these antibiotics, planting on nutrient agar containing the relevant antibiotic will allow only those cells that have been transfected and contain plasmid DNA to grow.

Term	Definition
The principle of DNA cloning	Most foreign DNA fragments cannot self-replicate in a cell and must therefore be joined (ligated) to a vector (virus or plasmid DNA) that can replicate autonomously. Each vector typically will join with a single fragment of foreign DNA. If a complex mixture of DNA fragments is used a population of recombinant DNA molecules is produced. This is then introduced into the host cells, each of which will typically contain only a single type of recombinant DNA. Identification of the cells that contain the DNA fragment of interest allows the purification of large amounts of that single recombinant DNA and hence the foreign DNA fragment.
Thermal induction	The change in growth and development of plants brought about by exposaure to a given temperature, usually applied to the process resulting in flowering of beiennial plants.
Three phases of transcription	Initiation → Elongation → Termination
Three-way cross	A three-way cross is a first generation hybrid between a single cross and an inbred line.
Threshold	A term used in studying effects of radiation. Below a certain dose, if there is a threshold, there is no measurable effect. No threshold exists for most genetic effects.
Ti plasmid	A plasmid of a tumor-inducing bacterium that integrates a segment of its DNA into the host chromosome of a plant.
Tiller	A branch arising from the base of a monocot plant, especially in the grass family.
Time of one-half response	Time required obtaining a change in moisture content from the original moisture content half of the way to equilibrium, usually expressed in hours.
Tissue	An aggregate of cell of similar structure performing similar functions.
Tissue culture	The growth and maintanance of cells from higher organisms *in vitro*, outside the tissue of which they are normally a part.

Term	Definition
Tolerance	The limit of difference between the results of two tests of the same attribute, beyond which the difference is considered a real one in the circumstances prescribed.
Trait	A synonym of character with respect to function and performance, but less so with respect to form.
Transduction	The transfer of genes for bacterial characters by means of a phase particle acting as a messenger boy.
Transformation	The heritable modification of the properties of one bacterial strain by an extract derived from cells of another strain; or the changing of the genotype of a microorganism by combining with a transforming principle supplied; this principle is DNA.
Transgenic organism	An organism containing certain genes from another species, produced for example by injecting foreign DAN into the nuclei of egg cells or early embryos.
Translocation	The exchange of parts of two non-homologus chromosomes, following breakage either spontaneous or induced.
Treated seed	Seeds to which only pesticides, dyes or other additives have been applied which have not resulted in a significant change in size, shape or addition to the weight of the original seed and which can still be tested according to standard methods.
Treatment	Any process, physical or chemical to which a seed lot is submitted.
Trihybrid	A hybrid involving three gene pairs such as A/a, B/b, C/c and the offspring from such a hybrid.
Triploid	Having three genomes or sets of chromosomes (3n).
Tropic levels	The producers and consumers in ecosystem can be arranged into several feeding groups each is known as tropic level. In any ecosystem, producers represents the first tropic level, secondary consumers at 2nd and 3rd tropic level and top carnivores represents the last level.

Term	Definition
Troposphere	The lowest layer of atmosphere in which man and other living organisms live is called troposphere.
Turgor pressure	The pressure caused by water in the vacuole, which pushes against the cytoplasm and cell wall. The turger pressure is balanced by the osmotic pressure when the cell is well supplied with water when the cell loses water, the turgor pressure decreases and more water can be enter the cell by osmosis.
Turgor	Stat of cell in which the cell wall is rigid, stretched by increase in volume of vacuole and protoplasm during absorption of water.
Turgor pressure	The pressure, which develops in a cell from time to time due to the osmotic diffusion of water, is called the turgor pressure.
Turner's syndrome	An abnormality in human beings; individuals are phenotypically females, but have rudimentary sexual organs and mammary glands. Such individuals have but one X chromosome and no Y with a total of 45 chromosomes instead of the normal 46.
Turning	The process of moving seed through the air from one bin or storage structure to another or back to the same storage.
Tyndall phenomenon	When a colloidal sol is kept in a glass vessel, preferably with flat parallel sides and a strong pencil of light is passed through it, the path of the light in the sol becomes luminous is called Tyndall phenomenon.
TZ	The term is applied to tetrazolium test for evaluating viability of seeds.
Uniformity	Refers to the uniformity of a seed lot; absence of apparently signicant variation in a seed lot.
Uracil	A nitrogen base, one of two pyrimidines found in RNA, but not DNA.
Urea	A soluble form of nitrogenous waste excreted by mammals and most adult amphibians.

Term	Definition
Urea cycle	A metabolic cycle taking place in the liver that prepares toxic ammonia for safe travel through the blood and then excretion by the kidney.
Ureter	A duct leading from the kidney to the urinary bladder.
Uric acid	Uric acid, the major nitrogenous waste product of uricotelic organisms, is also formed in other organisms from the breakdown of purine bases. Gout is caused by the deposition of excess uric acid crystals in the joints.
Uterus	A female reproductive organ where eggs are fertilized and or development of the young occur.
Vaccine	A harmless variant or derivative of a pathogen that stimulates a host's immune system to mount defenses against the pathogen.
Vagina	A thin walled chamber that forms the birth canal and is the repository for sperm during copulation.
Valence shell	The outermost energy shell of an atom, containing the valence electrons involved in the chemical reactions of that atom.
van der Waals Interactions	When two atoms approach one another closely, they create a non-specific weak attractive force that produces a van der Waals interaction.
Vapour pressure	Molecules constantly escape from liquids to form gases. The pressure exerted by this gas at equilibrium is termed as vapour pressure. It varies directly with the temperature.
Vapour-proof storage	Airtight sealed storage.
Variegation	Diversity in characters in the same organism, *e.g.,* alternating patches of green and white in leaves of some plants.
Variety	The term variety or cultivar denotes an assemblage of cultivated individuals, which are distinguished by any characters (morphological, physiological, chemical or others) significant for the purpose of agriculture, forestry or horticulture and which when reproduced (sexually or asexually) or reconstituted, retain their distinguishing features.

Term	Definition
Variety	A strain released for commercial cultivation by a variety release committee.
Vas deferens	The tube in the male reproductive system in which sperm travel from the epididymis to the urethra.
Vector	A plasmid or a viral DNA molecule into which either a cDNA sequence or a genomic DNA sequence is inserted.
Vegetative	A descriptive term referring to stem and leaf development in contrast to flower and seed development.
Ventical or upright storage	A storage structure whose height is greater than the width or diameter. Other terms used are silos, tanks and deep bins.
Ventilation front	The locus of all points of equal traverse time in the product being conditioned.
Viable (viability)	Alive, ability to live, grow and develop. A viable seed is one, which is capable of germinating under the proper circumstances. Such a viable seed may or may not be readily or immediately germinable. Dormant viable seeds may require lengthy specific treatments before they become immediately germinable.
Vibratory separator	It consists of a tilted deck surface, which is vibrated by an electromagnetic vibrator. Some of the difficult separations, which could not be made on other machines owning to minor differences in shapes and texture, could be made on this separator.
Vigour	Vigour is the sum total of all seed attributes which favours rapid and unform stand establishment in the field.
Virulent phase	A virus that kills the host bacterium. The contrasting type is temperate phase.
Virus	A small infectious agent required a host cell in which to reproduce. Its principal structure consists of DNA or RNA surrounded by a coat of protein; or a parasitic particle in plants and animals, some times causing disease, incapable of reproduction outside of the host cell.

Term	Definition
Vitamins	An organic molecule required in the diet in very small amounts; vitamins serve primarily as coenzymes or parts of coenzymes.
	Organic substances needed by the body in tiny amounts, the dietary absence of which can cause specific metabolic defects. These substances may in some instances be synthesized by humans but may not be synthesized in sufficient quantity to support normal health.
Volunteer plant	Plants from seeds of earlier crop or from accidentally planted seed are known as 'volunteer plant'.
	Unwanted plants growing from seed that remains in the field from a previous crop.
Water potential	The sum of osmotic potential and wall pressure denotes a net change in the chemical potential of water relative to that pure water at the same atmospheric pressure and temperature. This net change in the chemical potential is termed the water potential.
	In osmosis, the tendency for system to take up water from pure water, through a differentially permeable membrane.
Wavelength	The length of a wave of light. Different wavelengths have different colours and different levels of energy.
Weed	Any plant in a place where it is a nuisance might be considered a weed. The term is most often applied to non-cultivated plants that arise unwanted in cultivated areas, lawns, pastures or other areas used by man. Most weeds are prolific and persistent.
Wet bulb temperature	The temperature of the air as measured by an ordinary thermometer whose glass bulb is covered by a wet cloth or gauge. The temperature is recorded after the thermometer has been moved rapidly in the air.
Wilson's disease	A hereditary syndrome transmitted as a recessive trait in which liver proteins cause increased binding of copper.

Term	Definition
Working sample	A reduced sample taken from the submitted sample in the laboratory, on which one of the seed quality tests is made.
	Small portion of specified weight of submitted sample taken for specific seed test in STL by concerned officer is called as working sample.
Xanthophylls	A yellow-coloured compound $C_{40} H_{56} O_2$ found in plants.
Xo condition	Having an X chromosome but no Y chromosome, as the female in poultry and same other birds.
X-ray crystallography	The use of diffraction patterns produced by X-ray scattering from crystals to determine the 3-D structure of molecules.
Xylem	The tube shaped nonliving portion of the vascular system in plants that carries water and minerals from the roots to the rest of the plant.
Yeast	A unicellular fungus that lives in liquid or moist habitats, primarily reproducing asexually by simple cell division or by budding of a parent cell.
Yield	The net recoverable amount of the desired product from a reaction or process.
Zwitterions	Amino acids in solution at neutral pH are predominantly dipolar ions rather than unionized molecules are called zwitterions.
Zygote	The result of the fusion of male and female gametes; the individual that develops from such a fusion, usually diploid with 2n chromosomes.
Zygotic lethal	A lethal gene whose effect is in the embryo, larva or adult in contrast to the gametic lethal affecting a gamete.
Zymase	A mixture of numerous enzymes (at least 14).
Zymogen	An inactive precursor form of an enzyme.
	An inactive enzyme that becomes active only in the presence of its substrate.

6

REASONING / SHORT NOTES

Seed Technology

Seed technology as that discipline of study having to do with seed production, maintenance, quality and preservation.

Seed technology can be defined as the methods through which the genetic and physical characteristics of seeds could be improved. It involves such activities as variety development, evaluation and release, seed production, processing, storage and certification.

In a narrow sense seed technology can be described as a technique of seed production, seed processing, seed storage, seed testing and certification, seed marketing and distribution and the related research on these aspects too.

Types of Seed Dormancy

(i) **Primary dormancy**

Due to inherent properties of seed or due to unfavourable environmental conditions – 2 types.

    (a) **Exogenous /enforced/relative dormancy:** Due to absence of essential germination components like water, light, temperature etc.

    (b) **Endogenous/innate/true dormancy:** Due to inherent properties of seed.

(ii) **Secondary dormancy**

Induced by manipulation of physical and environmental factors.

(iii) **Physical dormancy**

Due to presence of growth inhibitors or absence of growth promoters. Presence of ABA and Auxins in low concentration inhibits the germination.

(iv) **Embryo dormancy**

Due to physiological immaturity of embryo.

Seed Dormancy Advantages

(1) Pre harvest- sprouting (vivipary) is checked in some varieties of certain crops, *e.g.*, groundnut.

(2) In adverse storage situation seed quality is maintained due to impermeability of seed coat, *e.g.*, cotton.

Methods of Breaking Seed Dormancy

(1) **Scarification:** Mechanical scarification, rubbing seed with sand paper, piercing, cutting acid scarification, seed soaking in sulphuric or nitric acid (1 to 60 minutes)

(2) **Stratification (temperature treatment)**

 (i) **Prechilling treatment:** Moistened seeds are kept at low temperature (5 – 10 °C) up to 7 days.

 (ii) **Preheating or drying treatment:** Seeds are dried at 30 to 35°C for 7 days with free air circulation.

 (iii) **How water treatment:** Seeds are soaked at 80°C for 1 to 5 minutes.

Physiological Dormancy

KnO_3 Seed treatment 0.2%

0.05% GA Seed treatment.

Isolation Distance Requirements for Various Agricultural Crops Seed Production.

S.No.	Crop	Isolation Distance (m)
1.	Paddy	3
2.	Whaet	3, 150 for India
3.	Barley	3, 150 for controlling smut disease
4.	Sorghum/Jowar	200
5.	Forage Sorghum	400
6.	Johnson grass (Sorghum halepense)	400

S.No.	Crop	Certified seeds	Foundation seeds
7.	Bajara	200	400
8.	Maize	200	400
9.	Red gram	100	200
10.	Gram	5	10
11.	Pea	5	10
12.	Black gram	5	10
13.	Green gram	5	10

Contd...

Contd...

S.No.	Crop	Certified seeds	Foundation seeds
14.	Lentil	5	10
15.	Cowpea	5	10
16.	French pean	5	10
17.	Hyacinth bean (Dolichos bean)	5	10
18.	Groundnut	3	3
19.	Rape and Mustard	50	100
20.	Linseed	25	50
21.	Sesame	50	100
22.	Caster	100	300
23.	Sunflower	200	400
24.	Soybean	3	3
25.	Niger	200	400
26.	Safflower	200	400
27.	Cotton	30	50
28.	Jute	30	50
29.	Mesta	100	200
30.	Sunhemp	100	200
31.	Berseem	100	400
32.	Lucerne	100	400
33.	Oats	150	150
34.	Guar	5	10
35.	Senji (Indian clover)	25	50
36.	Metha	200	400
37.	Sugar beet	1000	1600
38.	Sugarcane	5	5
39.	Tomato	25	50
40.	Brinjal	100	200
41.	Hot pepper and Sweet pepper	200	400
42.	Okra	200	400
43.	All types Cucurbits	500	1000
44.	Cauliflower	1000	1600
45.	Cabbage	1000	1600
46.	Knol-Khol	1000	1600
47.	Spinach (Palak)	1000	1600
48.	Lettuce	25	50
49.	Amaranthus	200	400
50.	Methi, Kasuri methi	5	10
51.	Garden beet	1000	1600
52.	Radish	1000	1600
53.	Turnip	1000	1600
54.	Carrot	800	1000
55.	Onion	500	1000
56.	Potato	5	5

Chemical Composition of Cytoplasm

Component	Content (%)
Carbohydrates	1.0
Proteins	10 – 20
Lipids	2.3
DNA	0.4
RNA	0.7
Inorganic matters	1.5
Other organic matters	0.4
Water	75 - 85

Functions of Water in Plants

Water is the most important component in the living matter. It is the medium for many biochemical reactions and extraction processes. Some of its functions are enumerated as below:

1. It is the medium in which many metabolic activities occur.

2. It is a major constituent of protoplasm

3. It is the solvent in which mineral nutrients enter a plant from the soil solution and transported to various parts of plant.

4. It is a reactant and product in a number of metabolic reactions.

5. In photosynthesis the hydrogen atom in the water molecule is incorporated into organic compounds and oxygen atoms are released as O_2.

6. It helps in the maintaining the turgidity to growing cells and thus maintains their form and structure.

7. Gain or loss of water from cells and tissues is responsible for variety of movements of plant parts.

8. The elongation phase of cell growth depends on absorption of water.

9. It is metabolic end product of respiration.

10. Plants absorb considerable amount of water and greater amounts of water are lost by plants then other substrance.

11. Water acts as the solvent in which reactions occurs, the medium for transport of enzymes, cofactors and coenzymes.

12. It acts substrate in hydrolytic reactions such as enzymatic breakdown of the polymers of starches, proteins, lipids and other complex molecules in to their basic uints.

13. Water provides turgor pressure for increase in cell volume and accounts for the 30 – 40% increase in volume upon initital hydration of the seed.

14. Hydration of the seed to levels above 16 – 18% moisture results in a rapid rise in mitochondrial activity and activation of polychrome.

15. Increase the rate of syntheis of proteins and ribonucleic acids.

Examples of Resistance to Individual Stresses

1. **Temperature:** Plants with few exceptions attain the temperature of the ambient environment they are poikilotherms. Because of this they must have some form of tolerance to temperature stress.

2. **Drought:** Terrestrial plants are normally turgid and thus resistance results from avoiding loss of turgor. Drought resistance should probably be divided into dehydration avoidance or postponement and dehydration tolerance.

3. **Irradiation:** Because of the penetrating nature of irradiation plants cannot escape it. Tolerance of irradiation stress depends upon the intensity and duration of the radiation and the amount of energy absorbed.

4. **Salts:** Plants that grow with root systems in soil of high salt content have low osmotic potentials as a result of an increased concentration of solutes and are salt tolerant. Some plants are resistant because they have mechanisms by which exclude salt or by which the salt is concentrated in vacuoles.

5. **Nutrient deficiency:** Tlerance of deficiency of a nutrient may depend on the ability of the roots to exude metabolites enabling them to obtain more of the nutrient from the soil as with iron or to use substitute ions such as sodium for potassium.

6. **Nutrient toxicity:** Much of the tolerance of toxicity is the result of the processes of exclusion at the root or by concentration in vacuoles of leaf cells. Either process prevents the toxic concentration from interfering with metabolism.

Steps Needed for Commercial Synthetic Seed Prodution

1. Production of enbryogenic tissue from transformed cells or tissues.

2. Large-scale production of synchronous somatic embryos.

3. Maturation of somatic embryos.

4. Non-toxic encapsulation process.

5. Artifical endosperm depending on species.

6. Storage capability of artifical seeds.

7. High frequency direct greenhouse field conversion depending on production requirements.

8. Low gentic and epigenetic variations.

9. Appropriate expression of engineered traits.

10. Artifical seed propagation.

11. Artifical seed propagation could potentially reduce the time needed to insert a desirable gene into a productive forest, as compared to using seed as the propagation method.

12. Production of large-scale embryogenic tissue from genetically engineered cells.

13. Concurrent plant regeneration, conformation of transformation and progeny testing.

14. Crygenic storage of potntial superior lines.

15. Scale-up production and maturation of somatic embryos.

16. Encapsulation of somatic embryos as artificial seeds.

17. Greenhouse establishment, growth and transplanting into the field or direct seedling.

18. The final stage will be the evaluation of production plantations for increased yield/performance due to engineered trait.

Stomatol Numbers, Epidermal Cell and Stomatol Index of Different Field Crops

Crop	Stomatol number	Epidermal cell	SI (%)
Bean	4	12	25.0
Cajanus	5	12	29.4
Cotton	6	32	15.8
Groundnut	9	36	20.0
Sorghum	8	20	28.6
Tomato	7	25	21.9

Capabilities of Different Plant to Fix Nitrogen

Host plant	Nitrogen fixed capability (kg/ha/yr)
Alfalfa	600
Alnus glutinosa	50
Casuarina equisetifolia	100
Casuarina littoralis	250

Host plant	Nitrogen fixed capability (kg/ha/yr)
Cowpea	100
Gliricidia sepium	13
Leucaena leucocphala	500
Myrica gale	10
Pea	120
Pueraria phaseoloides	650
Red clover	200
Soybean	100

Different Types of Proteins Present in Plant and Animal Tissues

Type of Protein	Nature / Specification
Albumins	The albumins are soluble in water and in dilute salt solutions. They are coagulated by exposure to heat.
Globulins	The globulins are insoluble or sparingly soluble in water and are soluble in dilute salt solutions. They are coagulated by exposure to heat.
Prolamines	The prolamines are insoluble in water but are souble in 70 to 80 per cent ethanol. They are called prolamine because they yield relatively large quantities of proline and ammonia on hydrolysis.
Glutalins	The glutalins are insoluble in water, alcohol and salts solutions but are soluble in weak acid of basic solutions.
Histones	The histones are soluble in water but are insoluble in dilute ammonia. They are rich in basic aminoacids, such as a ginine and lysine.
Conjugated proteins	Conjugated proteins are proteins, which contain non-amino acid components in addition to the aminoacids. The non-amino acids components or non-proteinaceous groups are called prosthetic groups. The comjugated are named in accordance with their associated prosthetic group. The conjugated proteins are divided into seven major types on the basis of nature of prosthetic groups, glycoproteins, lipoproteins, chromoproteins and metalloproteins.
Nuceoproteins	They are found in the nucleus comjugated with nucleic acids. They are called nucleoproteins as they give rise to simple proteins and nucleic acids on hydrolysis. They are weakley acidic and soluble in water.
Glycoproteins	They are found in cell membrane. They are proteins containing small amounts of carbohydrates as prosthetic groups and hence called glycoproteins.
Lipoproteins	They are found in cell memberane of nucleus, mitochondria, chloroplast and cell. They are comjugates of lipids and proteins and the prosthetic group of them are lipids. They are generally insoluble in water.

Contd...

Type of Protein	Nature / Specification
Chrmoproteins	They are found in flvoproteins, carotenoids, chlorphylls and hemoglobins. They posses pigment as their prospthetic groups and are cloured. The colour is because of metals such as Cu, and Va or metals with organic groups such as in Fe and Mg prophyrins.
Metalloproteins	They are found in respiratory enzymes, which require metals as activators.
Metaproteins	Metaproteins are insoluble in water and in dilute salt solution but soluble in acids and alkalies. These are produced by hydrolysis of simple proteins by alkalies or acids.
Coagulated proteins	Coagulated proteins are insoluble in water and are produced by the prolonged treatment with alcohol or heat. These are coagulated on heating.
Proteoses	Proteoses are soluble in water but are not coagulated by heat. These are produced by hydrolysis of metaproteins.
Peptones	Peptones are soluble in water and are not coagulated by heat. Peptones are produced by hydrolysis of proteoses by action of HCl, H_2SO_4 or certain enzymes.
Peptides	Peptides are soluble in water and are not coagulated by heat. These are produced by prolonged hydrolysis of simple proteins with HCl or H_2SO_4. They do not give biuret test.

Inlist of Biochemical Actions of Ethylene in Plants

1. RNA syntheis
2. Phenylalanine ammonia lyase
3. Swelling
4. Volume change
5. Inducement of alternative path of respiration
6. Malic enzyme
7. Peroxidase
8. Leakage of peroxidase
9. Release of alfa-amylase
10. Stabilization of enzyme activity
11. Ribosomes
12. Chloroplast syntheis
13. Chlorphyll synthesis
14. Chlorphyll degradation
15. Degreening

16. Carotenoid syntheis
17. Pigments formation
18. Formation of anthocyanins
19. ATPase stimulation
20. Mitochondria activation
21. Modification of ribosomes
22. Chitinase
23. Catalase
24. α–Amylase
25. Polygalacturonase
26. Pectinase
27. Glucanase
28. Cellulase
29. Enzymes activations
30. Protein loss
31. Protein synthesis

Environmental Sources of Stresses for Plants

Abiotic

1. Drought
2. Temperature
3. Air pollution
4. Allelochemicals
5. Nutrients
6. Radiation
7. Flooding
8. Pesticides
9. Toxins
10. Salts
11. Mechanical
12. Electrical
13. Magnetic
14. pH of soil solution

Biotic

1. Allelopathy
2. Competition
3. Diseases
4. Insects
5. Pests
6. Human activites
7. Lack of symbiosis

Steps Needed for Synthetic Commercial Seed Production

1. Production of embryogenic tissue from transformed cells or tissue.
2. Large-scale production of synchronous somatic embryos.
3. Maturation of somatic embryos.
4. Non-toxic encapusalation process.
5. Artificial endosperm depending on species.
6. Storgae cabaility of artificial seeds.
7. High frequency, direct green house field conversion, depending on production requirements.
8. Low genetic and epigenetic variation.
9. Appropriate expression of engineered trait.
10. Artificial seed propagation.
11. Artificial seed propagation could potentially reduce the time needed to insert a desired gene into a productive forest, as compared to suing seed as the propagation method.
12. Production of large embryogenic tissue genetically engineered cells.
13. Concurrent plant regeneration, confirmation of transformation and progeny testing.
14. Cryogenic storage of potential superior lines.
15. Scale-up production and maturation of somatic embryos.
16. Encapsulation of somatic embryos as artificial seeds.
17. Either greenhouse establishement, growth or transplanting into the field or direct seedling.
18. The final stage will be the evaluation of production plantations for increased yield/performance due to engineered trait.

- Stomata get closed when there is deficiency of water even though light and temperature conditions are favourable.

- In the case of hydroactive, control abscisic acid (ABA) plays an important role in the closure of stomata.

- Stomatal openings are essentially meant for an easy exchange of gases required for such important and indispensable physiological processes as photosynthesis and respiration. This type of anatomy of the leaf, however, has unavoidably led to the dreadful phenomenon of transpiration.

- Elements like sulphur, phosphorus and nitrogen are required in the formation of protein, which is an important constituent of protoplasm. These elements are therefore, known as protoplasmic elements.

- The highest rate of photosynthesis occurred in the orange-red light, the next in the green yellow light and lowest in violet blue light.

- The resolving power of human eye is 0.1 mm. The ordinary light microscope has a resolving power of about 0.2 mm. Ultraviolet light microscope can resolve upto 0.1 mm. The protoplast involves four parts; 1. the cytoplasm, 2. the vacuole, 3. a number of inclusions and 4. the nucleus.

- The mass of the cytoplasm is called the mesoplasm. The vacuole is separated from the cytoplasm by the vacuolar membrane called the tonoplast.

Enzyme and their Respective Reaction

Enzyme	Reaction
Glucosekinase	Glucose + ATP $\rightarrow$ glucose 6-P
Fructokinase	Fructose + ATP $\rightarrow$ fructose 1-P
Galactokinase	Galactose + ATP $\rightarrow$ galactose 1-P
Hexokinase	Hexose + ATP $\rightarrow$ Hexose-6-P
Triosekinase, Glucosolactonase	Glyceraldehydes + ATP $\rightarrow$ phosphoglyceraldehyde 6-phosphogluconolacton
6-Phosphogluconic dehydrogenase	6-phosphogluconate $\rightarrow$ ribulose-5-P
Phosphopentokinase	Ribulose 5-P + ATP $\rightarrow$ ribulose-1, 5-diphosphate
Enolase	2-phosphoglycerate $\rightarrow$ phosphoenolpyruvate
Pyruvic kinase	Phosphoenolpyruvate $\rightarrow$ pyruvate + ATP
Carboxylase	Pyruvate $\rightarrow$ acetaldehyde + CO
Phosphoglyceric kinase	1,3-diphosphoglycerate + ADP $\rightarrow$ 3-phosphoglycerate + ATP

Substrate and their Respective Respiratory Quotients

Substrate	Respiratory quotient
Carbohydrates with some anaerobic respiration, Carbohydrates synthesized from organic acids, Organic acids	> 1.0
Carbohydrates	1.0
Proteins with NH_3 formation	0.99
Proteins with amide formation	0.8
Fats, *e.g.*, tripalmitin	0.7
Fats with associated carbohydrate synthesis	0.5
Carbohydrates with associated organic acid synthesis	0.3

The *m*RNA Genetic Code Dictionary

Codons	Coded for Amino acid	Codons	Coded for Amino acid
UUU	Phenylalanine	UAU	Tyrosine
UUC	Phenylalanine	UAC	Tyrosine
UUA	Leucine	UAA	Ochre (Terminator)
UUG	Leucine	UAG	Amber (Terminator)
CUU	Leucine	CAU	Histidine
CUC	Leucine	CAC	Histidine
CUA	Leucine	CAA	Glutamine
CUG	Leucine	CAG	Glutamine
AUU	Isoleucine	AUU	Aspargine
AUC	Isoleucine	AAC	Aspargine
AUA	Isoleucine	AAA	Lysine
AUG	Methionine (Initiator)	AAG	Lysine
GUU	Valine	GAU	Aspartic acid
GUC	Valine	GAC	Aspartic acid
GUA	Valine	GAA	Glutamic acid
GUG	Valine	GAG	Glutamic acid
UCU	Serine	UGC	Cystine
UCC	Serine	UGC	Cystine
UCA	Serine	UGA	Opal (Terminator)
UCG	Serine	UGG	Tryptophan
CCU	Proline	CGU	Arginine
CCC	Proline	CGC	Arginine
CCA	Proline	CGA	Arginine
CCG	Proline	CGC	Arginine
ACU	Threonine	AGU	Serine
ACC	Threonine	AGC	Serine
ACA	Threonine	AGA	Arginine
ACG	Threonine	AGG	Arginine
GCU	Alanine	GGU	Glycine
GCC	Alanine	GGC	Glycine
GCA	Alanine	GGA	Glycine
GCG	Alanine	GGG	Glycine

7

DIFFERENCES / COMPARING

Difference between guard cells in illuminated and dark conditions

Illuminated Guard Cells	Dark Guard Cells
1. Respiratory CO_2 contained in the intercellular spaces is used up by the mesophyll in photosynthesis.	1. Respiratory CO_2 accumulates in the intercellular spaces.
2. pH of the guard cells rises.	2. pH of the guard cells falls.
3. The decrease in activity favours hydrolysis of starch into sugars.	3. The increase in acidity favours conversion of sugar into starch.
4. The O. P. of the cell-sap of the guard cells increases.	4. The O.P. of the sap of guard cells decreases.
5. Water enters the guard cells, and their turgor pressure and volume increases.	5. Water leaves the guard cells and their turgor pressure and volume decreases.
6. The guard cells change their shape and stomatal apertures widen.	6. The guard cells change their shape and stomatal apertures are narrowed.

Difference between different chemical solutions

Sach's Solution, 1860		Knop's Solution, 1865	
Salt	g/l	Salt	g/l
KHO_3	1.0	$Ca(NO_3)_2.4H_2O$	0.8
$Ca_3(PO_4)_2$	0.5	KHO_3	0.2
$MgSO_4.7H_2O$	0.5	KH_2PO_4	0.2
$CaSO_4$	0.5	$MgSO_4.7H_2O$	0.2
$ZnCl$	0.25	$FePO_4$	Trace
$FeSO_4$	Trace		

Shive's Solution, 1915		Hoagland's Solution, 1920	
Salt	g/l	Salt	g/l
$Ca(NO_3)_2.4H_2O$	1.06	$Ca(NO_3)_2.4H_2O$	1.18
KH_2PO_4	0.31	KHO_3	0.51

Contd...

Contd...

Shive's Solution, 1915		Hoaglond's Solution, 1920	
MgSO$_4$.7H$_2$O	0.55	KH$_2$PO$_4$	0.14
(NH$_4$)$_2$ SO$_4$	0.09	MgSO$_4$.7H$_2$O	0.49
FeSO$_4$. 7H$_2$O + Minor elements	0.005	Ferric tartarate+ Minor elements	0.005

Difference between C$_3$ and C$_4$ plants

C$_3$ Plants	C$_4$ Plants
1. Plants, which have Calvin cycle in all the green cells of the leaf.	1. Plants, which have Hatch-Slack cycle in the mesophyll and Calvin cycle in the cells of the bundle sheath.
2. There is only one CO$_2$ acceptor, the Ribulose 1, 5 diphosphate, which occurs in all the green cells of the plant.	2. There are two CO$_2$ acceptors phosphoenolpyruvate (PEP) in the mesophyll and RuDP in the cell of the bundle sheath.
3. The first stable product of photosynthesis is a 3-carbon compound, phosphoglyceric acid.	3. The first stable product is a 4-carbon compound, oxaloacetic acid.
4. "Kranz" anatomy is absent. The chloroplasts in all the green cells of the leaves are of the normal type. They have well-defined grana and have both the photosystems I and II.	4. The leaves have "Kranz" type of anatomy. The vascular bundles are surrounded by bundle sheath cell. The chloroplasts are dimorphic. The mesophylls have chloroplasts similar to those of C$_3$ plants. The cells of the bundle sheath have very large chloroplasts. They lack grana and contain starch grains. They are centripetally arranged within the cells of the bundle sheath. These chloroplasts lack photosystem II, and therefore, have to depend upon the chloroplasts of the mesophyll for a supply of NADPH$_2$$^+$.
5. Ribulose diphosphate and the enzymes of the Calvin cycle are present within all the green cells and therefore, Calvin cycle occurs in the mesophyll.	5. Ribulose diphosphate and the enzymes of the Calvin cycle are present in the bundle sheath cells only whereas the mesophyll cells contain phosphoenolpyruvic acid and the enzymes of the Hatch-Slack cycle.
6. The optimum temperature ranges between 10 – 25°C.	6. The optimum temperature range between 30 – 45°C.
7. The maximum light intensity for photosynthesis is 1000 – 1200 foot candles, which is the saturation intensity.	7. C$_4$ plants can photosynthesize at a much higher light intensity, and even full sunlight is not the saturation intensity.
8. Because of lesser affinity of RuDP for CO$_2$ the Calvin cycle can reduce the CO$_2$ Concentration around the plants to only 50 ppm.	8. Because of stronger affinity of PEP for CO$_2$ the Hatch-slack cycle can reduce the CO$_2$ concentration to even less than 10 ppm.
9. C$_3$ plants are less efficient in photosynthesis. The net rate of the process in only 15–35 mg of CO$_2$ per dm^2 of leaf area per hour.	9. C$_4$ plants are more efficient in photosynthesis. The net rate of the process is 40 - 80 mg of CO$_2$ per dm^2 of the leaf area per hour.

C_3 Plants	C_4 Plants
10. Oxygen has an inhibitory effect on photosynthesis.	10. Oxygen doesn't have any inhibitory effect on the process.
11. Photorespiration occurs which further reduces the photosynthetic yield.	11. Photorespiration is absent and, therefore, the efficiency of photosynthesis is further increased.
12. The optimum temperature for the process is 10 – 25°C.	12. In these plants, it is 30 – 40°C. So these are warm climate plants. At this temperature, the rate of photosynthesis is double than that of C_3plants.
13. Oxygen present in air (=21% O_2) markedly inhibits the photosynthetic process as compared to an external atmosphere containing no oxygen.	13. The process of photosynthesis is not inhibited in air as compared to an external atmosphere containing no oxygen.
14. For synthesis of one glucose molecule 18 ATP are required.	14. In this process 30 ATP are required for the synthesis of one glucose molecule.

Difference between photosynthesis found in bacteria and green plants

Bacteria	Green plants
1. Bacteria have no distinct chloroplast.	Chlorplast is well developed.
2. Bacteria absorb light of longer wavelength (800–900 nm or infrared).	These absorb light of relatively shorter wavelength (450-700 nm).
3. P_{890} is the reaction center.	P_{680} and P_{700} are two reaction centers.
4. Chlorophyll a is absent and bacteriochlorphyll takes its function.	Chlorophyll a is present which converts radation to chemical energy.
5. The carotenoids are open chain aliphatic type.	Carotenoids are bicyclic.
6. Oxygen is not evolved during the process (anoxygenic photosynthesis).	Oxygen is evolved as a by-product (oxygenic photosynthesis).
7. Water does not serve as source of reducing power (electron donor).	Water serves as a source of reducing power.
8. Bacteria can use CO_2 as well as organic compounds as source of carbon.	Only CO_2 is the source of carbon.
9. The photoreductant is $NADH_2$	The photoreductant is $NADPH_2$
10. The process occurs in presence of light but in absence of O_2	Both are present during the process.
11. Emerson effect is not found.	Emerson effect if found.
12. Cyclic photophosphorylation is dominant.	Non-cyclic photophosphorylation is dominant.
13. Plastocyanin is absent.	Plastocyanin is present.
14. Lycopene is found.	Lycopene is not found in the chloroplast of higher plants.

Difference between prokaryotic and eukaryotic cells

S.No.	Characters	Prokaryotic cell	Eukaryotic cell
1.	Capsule	When present, composed of mucopolysaccharides	Absent
2.	Cell wall	Non-cellulosic; composed of amino-sugars and muramic acid	Absent in animal cells; present in plant cell cellulosic
3.	Chromosomes	Single	Multiple
4.	Division	Amitosis	Mitosis or meiosis
5.	DNA	Nacke	Combined with protein
6.	Endomembranes (ER, Golgi body, Mitochondria, Chloroplast lysosomes)	Absent	Present
7.	Nuclear envelope	Absent	Present
8.	Nucleolus	Absent	Present
9.	Nucleoplasm	Not differentiated from cytoplasm	Differentiated from cytoplasm by nuclear envelope
10.	Photosynthesis	In plasma membrane or thylakoids vesicles	In chloroplast
11.	Plasma memberane	Present	Present
12.	Respiratory enzymes	Enzymes in the plasma membrane	In mitochondria
13.	Ribosomes	Present: 70S (50S + 30S)	Present: 80S (60S + 40S)

Difference between active and passive absorption

S. No.	Active absorption	Passive absorption
1.	Movement of ions across membrane takes place according to concentration gradient.	Movement of ions across membrane takes place against concentration gradient.
2.	No energy is required.	Expenditure of energy is required.
3.	No carrier is necessary for this process.	A specific carrier is necessary in this process.
4.	No enzymes are involved.	Enzymes may be involved.
5.	The active absorption of water takes place due to activity of root and root hairs.	It occurs due to activity of upper part of the plant, such as shoot and leaves.
6.	The absorption of water occurs by the osmotic and non-osmotic processes.	The water is absorbed due to the active transpiration in aerial parts.
7.	The roots hairs have high DPD as compared to soil solutions and therefore, water is taken in.	The absorption occurs due to tension created in xylem sap by transpiration pull. Thus, water is sucked in by the tension.
8.	In the movement of water, the living part of protoplasts (symplast) is involved.	The movement of water is thorough free spaces or apoplast of root and it may include cell wall and intercellular spaces.
9.	The rate of absorption depends upon DPD or difference in osmotic concentrations between the two.	The rate of absorption depends upon transpiration.
10.	In non-osmotic type of absorption, respiratory energy is utilized.	Energy is never required.

Difference between evaporation and transpiration

Evaporation	Transpiration
1. It is just a pysical phenomenon.	It is vital and physiological phenomenon.
2. Guard cells are not involved in it.	It is regulated by the activity of gurad cells.
3. Any kind of pressures is not involved.	Transpiration involves different types of pressures.
4. No guard cells or stomata are involved	Guard cells regulate the surface area of transpiration.
5. It may take place from any surface.	It is process of the living cells.

Difference between traspiration and guttation

Transpiration	Guttation
1. This takes place in day time.	This takes place in night
2. Water is lost as vapours.	Water is lost as liquid
3. Water is transpired through stomata, lenticel or cuticle.	Water lost in guttation is rich in inerals, etc. through hydathode.
4. It is regulated phenomenon. Temperature of leaf and plant is decreased.	It is uncontrolled phenomenon. No such effect occurs.
5. The transpired water is pure.	Guttated water contins dissolved salts and sugar.
6. It lower down the temperature of the surface.	It lacks such a relationship.

Difference between DNA and RNA

Sr. No.	DNA	RNA
1	Sugar is 2-deoxy-D-ribose in which at position 2, oxygen is absent.	Sugar is D-ribose.
2.	Instead of uracil, thymine is present.	Nitrogenous bases-adenine, Guanine, Cytosine and Uracil.
3.	It is arranged according to the model proposed by Watson and Crick.	It is single stranded and without attaining any definite shape.
4.	Commonly found in the chromatids of nucleus.	Commonly found in the cytoplasm.

Difference between cyclic and non-cyclic photophosphorylation

Sr.No.	Cyclic photophosphorylation	Non-cyclic photophosphorylation
1.	In the process only PS I is functional.	Both PS I and II are functional.
2.	Electron moves in a closed circle. Electron freed from chlorphyll after excitation to acceptors results to chlorophyll.	In the process, water is the ultimate source of electrons and $NADP^+$ is the final acceptor.

Contd...

Contd...

Sr.No.	Cyclic photophosphorylation	Non-cyclic photophosphorylation
3.	Reduced NADP ($NADPH_2$) is not formed and assimilation of CO is slowed down.	$NADPH_2$ is formed which is used in assimilation of carbon dioxide.
4.	Oxygen is not evolved.	Oxygen as by product is evolved.
5.	The system is found dominantly in Photosynthetic bacteria.	The system is dominant is green plants.
6.	The process is not inhibited by DCMU.	The process is stopped by use of DCMU.

Difference between dark respiration and photorespiration

Sr.No.	Dark respiration	Photorespiration
1.	Respiratory substrate may be carbohydrate, fat or protein.	The substrate is glycolate.
2.	The substrate may be recently formed or a stored one.	The substrate is always recently formed.
3.	The process occurs in cytosol and mitochondria.	It occurs in between chloroplast, cytosol, peroxisome and mitochondria.
4.	H_2O_2 is not formed during the process	H_2O_2 is formed
5.	In the process, several ATP molecules are produced.	ATP molecules are not formed.
6.	NAD is reduced to $NADH_2$.	Here it is reverse, i.e., $NADH_2$ is oxidized into NAD.
7.	Transamination reaction does not occur.	Such reactions are involved in the process.
8.	The process is dependent on O_2 concentration only to a limited extent.	It shows a close positive correlation with O_2 concentration.
9.	Ammonia is not formed.	Ammonia is formed in the mitochondria and the peroxisome.
10.	The process is not so sensitive as to rise in temperature.	Its rate is highly accelerated in between 25 to 35°C.
11.	The process is found in all living cells.	It is found only in green cells.
12.	It is found in dark as well as in light both.	It is found only in presence of light

Comparison of xylem vessels and phloem sieve tubes

Sr. No.	Vessels	Sieve tubes
A.	**Anatomical comparisons**	
1.	Vessels have open ends.	Sieve plants are present at the ends of sieve tubes.
2.	Vessels are wider.	They are very narrow.
3.	They have thick rigid lignified walls.	They have thin and more extensible cellulosic walls.
B.	**Physiological comparisons**	
4.	They are dead when mature but functional.	They are alive when mature and functional.

Sr. No.	Vessels	Sieve tubes
5.	They are permeable to both solute and solvent.	They are semipermeable.
6.	They have low sap concentration.	They have high sap concentration (O.P. = 15 to 34 atm).
7.	They do not have turgor pressure.	They turgid cells have high turgor pressure
8.	Tey are partially collapsed when functioning.	They are distended by pressure when functioning
9.	They absorb water or air when cut (except when root pressure functions).	They excude cell sap when cut.
10.	They translocate both solute and solvent.	They translocate only solute.
11.	Their translocation speed is up to 75 cm/min.	Their translocation speed is up to 5 cm/min (maximum).

Difference between respiration and photosynthesis

Sr.No.	Respiration	Photosynthesis
1.	Oxygen is absorbed in the process.	Oxygen is liberated in the process.
2.	Carbon dioxide is evolved as a result of oxidation of carbon containing compounds.	Carbon dioxide is absorbed and is fixed inside to form carbon containing compounds.
3.	The process occurs day and night.	Process occurs only in presence of light.
4.	Light is not essential for the process.	Light is essential for the process.
5.	During the process, potential energy is converted into kinetic energy.	During the process, radiant energy (light energy) is converted into potential energy.
6.	Raw materials used are glucose and oxygen.	Raw materials used are CO_2 and water.
7.	The presence of chlorophyll is not necessary.	Presence of chlorophyll is necessary for photosynthesis.
8.	Energy is released during the process hence it is an exothermic process.	Energy is stored during the process hence it is an endothermic process.
9.	Due to repiration the plant suffers with the loss of weight.	By the process, the weight is gained.
10.	It is a catabolic process and includes the destruction of stored food.	It is an anabolic process and includes the manufacture of food.
11.	The process includes dehydrolysis and decarboxylation.	It includes the process like hydrolysis and carboxylation.
12.	During the breakdown of glucose molecule, 38 ATP molecules are formed.	During the synthesis of one glucose molecule, 18 dATP molecules are utilized.

Contd...

Differenc between photorespiration and true respiration

Photorepiration	*True respiration*
A. The effect of oxygen concentration	
1. It increases steadily with oxygen concentration all the way from zero to one hund-red per cent.	It is saturated at oxygen concentrations as low as two or three per cent and virtually no increase in rate are observed above these concentrations.
B. Requirement for the photosynthetic apparatus	
2. It is intimately linked to photosynthesis and occurs only in cells, which are capable of photosynthesis.	It occurs in any living cell.
C. The nature of the substrate	
3. The nature of the substrate is different.	The nature of the substrate is different.
D. Energy relationship	
4. Biochemistry of photorespiration differs fundamentally from that of the true respiration. Photorespiration, however, does not lead to a net production of ATP or reduced nucleotides. Glycolet arises as an unwanted by product of photosynthesis. The bulk of the carbon of Glycolet, after oxidation, is converted to carbohydrate, but the remainder is converted to CO_2, which is released to CO_2 of photorespiration.	Biochemistry of true repiration differs dundamentally from that of the photorespiration. Respiration brings about the oxidation of its substrates to yield CO_2 and energy, mainly in the form of ATP and reduced pyridine nucleotides.

Difference between aerobic and anaerobic respiration

Sr. No.	*Aerobic respiration*	*Anaerobic respiration*
1.	It is common to all plants.	It is of rare occurrence.
2.	It goes on throughout the life.	It occurs for a temporary phase of life.
3.	Energy is liberated in larger quantity. In total, 38 ATP molecules/glucose molecule are formed.	Energy is liberated in lesser quantity. Only 2 ATP molecules are formed.
4.	The process is not toxic to plants.	It is toxic to plants.
5.	Oxygen is utilized during the process	It occurs in absence of oxygen.
6.	The carbohydrates are oxidized completely and are broken into CO_2 and water.	The carbohydrates are oxidized incompletely and ethyl alcohol and carbon dioxide are formed.
7.	The end-products are carbon dioxide and water.	The end-products are ethyl alcohol and carbon dioxide.
8.	The process takes place partly (*glycolysis*) in the cytosol and partly (*Krebs cycle*) inside the mitochondria.	The process occurs only in the cytosol.

Difference between oxidative and photophosphorylation

Sr. No.	Oxidative phosphorylation	Photophosphorylation
1.	It occurs during respiration.	It occurs during photosynthesis.
2.	In general it is found inside the mitochondria.	It takes place within the chloroplast.
3.	The process occurs on the inner memberne of cristal.	It occurs in the thylakoid membrane.
4.	Molecular oxygen is needed during terminal oxidation.	Molecular oxygen is not required.
5.	The energy released during electron transfer due to oxidation-reduction reaction is used during ATP formation.	The source of energy for conversion of ATP from ADP and Pi is external (light).
6.	The process takes place in elecron transport system involving cytochromes.	Pigment system I and II are involved during the process.
7.	The ATP molecules are released in the cytoplasm available for different metabolic reactions.	The produced ATP molecules are used up for CO_2 assimilation in the dark reaction of photosynthesis.

Difference between Z-DNA and B-DNA

Sr. No.	Z – DNA	B - DNA
1.	It shows left handed helical sense.	It shows right handed helical sense.
2.	The phosphate backbone follows a zigzag course.	The phosphate backbone is regular.
3.	The sugar molecules show alternating orientation so that repeating unit is a dinucleotide.	Orientation of sugar molecules is not alternating and the repeating unit is a mononucleotide.
4.	One complete turn of DNA has twelve base pairs or six repeating dinucleotide units.	One complete turn has only ten pairs or ten repeat units.
5.	The angle of twist per repeat unit is 60°.	The angle of twist per repeat unit is 34°.
6.	The angle of one complete turn is 45 Å (*i.e.* helix pitch).	The length of one complete turn is 34 Å (*i.e.* helix pitch).
7.	The diameter of DNA molecule is 18 Å.	The diameter of DNA molecule is 20 Å.
8.	The distance in rise between base pairs is 3.7 Å.	The distance in rise between base pairs is 3.4Å.
9.	The base pair tilt is 7°	The base pair tilt is 6°.
10.	The distance of P from axis of - dGpC is 8.0 Å dCpG is 6.9 3 Å	The distance of P from axis is 9.0 Å in both cases.

Difference between two types of ribosomes

Sr. No.	Particulars	70 S ribosome	80 S ribosome
1.	Occurrence	In prokaryotic cell and in chloroplast and mitochondria of eukaryotic cell.	Incytoplasm and nucleus of eukaryotic cell.
2.	Sedimentation coefficient	Average 70 S	Average 80 S
3.	Mol. Wt.	3 million	4-5 million
4.	Subunits	Small-30 S and large-50 S	Small-40 S and large-60 S
5.	rRNAs	3 molecules of rRNAs. 16 R rRNA in 30 S subunit and 5 S and 23 S rRNAs in 50 S subunit.	4 molecules of rRNAs. 18 S rRNA in 40 S subunit and 5 S, 5.8 and 28 S rRNAs in 60 S subunit.
6.	Mol. Wt. of rRNAs	5 S rRNA-40, 000, 16 S rRNA-5, 50,000 and 23 S rRNA-11, 00,000.	5 S rRNA-39, 000, 5.8 S rRNA-51, 000, 18 S rRNA-7,00,000 and 28 S rRNA-1, 700, 000.
7.	Protein molecules	21 molecules (from S_1-S_{21}) in 30 S subunit and 34 molecules (from L_1–L_{34}) in 50 S subunit.	33 molecules (from S_1 - S_{33}) in 40 S subunit and 49 molecules (from L_1 – L_{49}) in 60 S subunit
8.	Average mol. Wt. of proteins	11,800	21, 000
9.	RRNA protein ratio	2:1 (i.e. RNA rich)	1:1 (i.e. protein rich)
10.	First critical Mg ions level	0.5 mole Mg ions per mole of phosphorus.	0.3 – 0.1 mole Mg ions per mole of phosphorus.

Difference between osmosis and diffusion

Sr. No.	Osmosis	Diffusion
1.	It is special type of diffusion in which a semipermeable membrane in between the two solutions is required.	Presence of semipermeable membrane is not required.
2.	Osmosis occurs in between liquid medium.	Diffusion may occur in any medium, it may be between solid, liquid or gas.
3.	In osmosis, diffusion of only solvent molecules from low concentration of solution to higher concentration of solution takes place.	In diffusion, a net downward movement of a given substance from higher concentration to lesser concentration is found.
4.	Movement of water through a differentially permeable membrane from an area of their higher concentration to an area of lower concentration.	The process in which the movement of particles of solid, liquid or gases from an area of their higher concentration to an area of lower concentration.
5.	It is a special type of diffusion in which a semipermeable membrane in between the two solutions is required.	Presence of semipermeable membrane is not required.

Contd...

Contd...

Sr. No.	Osmosis	Diffusion
6.	Osmosis occurs in between liquid	Diffusion may occur in any medium, it may be between solid, liquid or gas.
7.	In osmosis, diffusion of only solvent molecules from low concentration of solution to higher low concentration of solution takes place.	In diffusion, a net dawn ward movement of a given substance from higher concentration to lesser concentration is found.

Difference between active absorption and passive absorption

	Active absorption	Passive absorption
1	The active absorption of water takes place due to the activity of root and root hairs.	1. It occurs due to activity of upper part of the plant such as shoot and leaves.
2.	The absorption of water occurs by the osmotic and non-osmotic processes.	2. The water is absorbed due to the active transpiration in aerial parts.
3.	The root hairs have high DPD as compared to soil solution and, therefore, water is taken in.	3. The absorption occurs due to tension created in xylem sap by transpiration pull. Thus water is sucked in by the tension.
4.	In the movement of water, the living parts of protoplasts (symplast) is involved.	4. The movement of water is through free spaces or apoplast of root and it may include cell wall and intercellular spaces.
5.	The rate of absorption depends upon DPD or difference in osmotic concentrations between the two.	5. The rate of absorption depends upon transpiration.
6.	In non-osmotic type of absorption, respiratory energy is utilized.	6. Energy is never required.

Difference between transpiration and evaporation

	Transpiration	Evaporation
1.	It is a modified phenomenon found in plants.	1. It is a physical process and found taking place on any free surface.
2.	It is regulated by the activity of guard cells.	2. No such mechanism found in evaporation.
3.	In the process, only living cells exposed to the atmosphere are involved.	3. It can occur from both living and non-living surfaces.
4.	This involves different type of pressures such as vapor pressure, diffusion pressure, osmotic pressure etc.	4. In the process, no such processes are involved.
5.	It helps in keeping the surface of leaf and young stem wet to protect from sun burning.	5. It causes dryness of the free surface.

Difference between transpiration and evaporation

Transpiration	Guttation
1. It occurs during day time.	1. It usually occurs in the night.
2. The water is given out in the form of vapour.	2. The water given out in the form of liquid.
3. The transpired water is pure.	3. Guttated water contains dissolved salts and sugar.
4. It takes place through stomata, lenticels or cuticle.	4. It occurs through a special structure called hydathode found only on leaf tips or margin.
5. It is controlled process.	5. It is an uncontrolled process.
6. It lowers dawn the temperature of the surface.	6. It lacks such a relationship.

Difference between guard cells in light and guard cells in dark

Guard cells in light	Guard cells in dark
1. CO_2 is consumed in photosynthesis.	1. CO_2 is released from respiration accumulates in the intercellular spaces. There is no photosynthesis.
2. pH rises, *i.e.,* guard cell show basic reaction.	2. pH fails, *i.e.,* guard cell show acidic reaction.
3. An increase in pH favours hydrolysis of starch into sugars.	3. A decrease in pH favours formation of starch from soluble sugars present in the cell.
4. Sugars increase osmotic pressure of cell sap of guard cells.	4. Due to presence of starch osmotic pressure of cells falls.
5. Water enters the guard cells due to raise in osmotic pressure and turgor pressure and volume of guard cells increases.	5. Water moves out of guard cells to subsidiary. cells and osmotic pressure, turgor pressure and volume of guard cells decreases.
6. Guard cells become turgid and change their shape.	6. Guard cell become flaccid due to loss of water and changes their shape.
7. As result stomata opens.	7. As result stomata closes.

Difference between chemosynthesis and photosynthesis

Chemosynthesis	Photosynthesis
1. It occurs only in colourless aerobic bacteria.	1. The process occurs in green plants including green bacteria.
2. During the process, CO_2 is reduced into carbohydrates without light and chlorophyll.	2. CO_2 and H_2O are converted into carbohydrates in presence of light and chlorophyll.
3. In the process, chemical energy released during oxidation of organic substances is used up to synthesize carbohydrates.	3. Light energy is converted into chemical energy and stored in the carbohydrates.

Contd...

Contd...

Chemosynthesis	Photosynthesis
4. No pigment molecule is involved in the process.	4. Several pigments are involved to absorb light energy to perform photochemical act.
5. Oxygen is not involved.	5. Oxygen is involved as a by-product.
6. Photophosphorylation does not take place.	6. During the process, photophosphorylation is found.

Difference between role of Mg and Fe

Role of Mg	Role of Fe
1. It is a constituent of chlorophyll and, therefore, essential for the formation of this pigment.	1. It is a constituent of cytochrome, ferredoxin, catalase, peroxidase, fluorochrome, hematin, globoids of aleurone grains etc.
2. It acts as a phosphorus carrier in the plant particularly in connection with the formation of seeds of high oil contents which contain compound lecithin.	2. It acts as a catalyst and electron carrier in respiration and also closely related with chlorophyll synthesis.
3. It is readily mobile and when its deficiency occurs, it is apparently transferred from older to younger tissues where it can be reutilized in growth processes. As a result, deficiency symptoms develop first on older leaves.	3. It is relatively immobile in plant tissues and its mobility is affected by several factors such as the presence of Mg, K deficiency, high P and high light intensity. Lack of mobility accounts for Fe deficiency first developing in younger leaves.
4. Mg is essential for the synthesis of fats and metabolism of carbohydrates and phosphorus.	4. Fe acts as co-factor for the metabolism of fats, proteins and carbohydrates.

Difference between photosynthesis in bacteria and green plants

Photosynthesis in bacteria	Photosynthesis in green plants
1. Bacteria have no distinct chloroplasts.	1. Chloroplast is well developed.
2. Bacteria absorb light of longer wavelength (800-900 nm or infrared).	2. These absorb light of relatively shorter wavelength (450-700 nm).
3. P_{890} is the reaction centre.	3. P_{680} and P_{700} are the reaction centres.
4. Chlorophyll a is absent and bacteriochlorophyll takes its function.	4. Chlorophyll a is present ant it converts radiant to chemical energy.
5. The carotenoids are open chain aliphatic type.	5. Carotenoids are bycyclic.
6. Oxygen is not evolved during the process (Anoxygenic photosynthesis).	6. Oxygen is evolved as a by-product (oxygenic photosynthesis).
7. Water does not serve as a source of reducing power (electron donour).	7. Water serves as a source of reducing power.
8. Bacteria can use CO_2 as well as organic compounds as source of carbon.	8. Only CO_2 is the source of carbon.

Contd...

Contd...

Photosynthesis in bacteria	Photosynthesis in green plants
9. The photoreductant is $NADH_2$.	9. The photoreductant is $NADPH_2$.
10. The process occurs in presence of light but in absent of O_2.	10. Both are present during process.
11. Emersion effect is not found.	11. Emersion effect is found.
12. Cyclic photophosphorylation is dominant.	12. Non-cyclic photophosphorylation is dominant.
13. Plastocyanin is absent.	13. Plastocyanin is present.
14. Lycopene is found.	14. Lycopene is not found in the chloroplasts of the higher plants.

Difference between macro-elements and micro-elements

Macro-elements	Micro-elements
1. The essential nutrients required in higher concentration for the normal growth of the crop.	1. The essential nutrients required in low concentration for the normal growth of the crop.
2. C, H, O, N, P, K, Ca, Mg and S.	2. Fe, Mn, Cu, Zn, Mo, B and Cl

Difference between dark respiration and photorespiration

Dark respiration	Photo-respiration
1. Respiratory substrate may be carbohydrate, fat or proteins.	1. The substrate is glycolate.
2. The substrate may be recently formed or a stored one.	2. The substrate is always recently formed.
3. The process occurs in cytosol and mitochondria.	3. It occurs in between chloroplast, cytosol, peroxysome and mitochondria.
4. H_2O_2 is not formed during the process.	4. It is formed.
5. In the process, several ATP molecules are produced.	5. ATP molecules are not formed.
6. NAD is reduced to $NADH_2$.	6. Here it is reserve, *i.e.,* $NADH_2$ is oxidized to NAD.
7. Transammination reaction does not occur.	7. Such reactions are involved in the process.
8. The process is dependent on O_2 concentration only to a limited extent	8. It shows a close positive correlation with O_2 concentration.
9. Ammonia is not formed.	9. Ammonia is formed in the mitochondria and the peroxisome. .
10. The process is not sensitive as to rise in temperature.	10. Its rate is highly accelerated in between 25 to 30°C.
11. The process is found in all living cells.	11. It is found only in green cells.
12. It is found in dark as well as lights both.	12. It is found only in presence of light.

Difference between inorganic catalysts and enzymes

Inorganic catalysts	*Enzymes*
1. These are small molecules or simple mineral ions.	1. These are made up of proteins with complex three dimensional structures.
2. These can catalyze diverse reactions.	2. These catalyze only specific reactions of a single or only a few substances.
3. These are not regulated by any regulator molecule.	3. These can be regulated by specific molecules which can change their confirmation and activity.
4. These are less sensitive to pH and temperature.	4. Enzymes are more sensitive to changes in PH and temperature.

Difference between C_3 plants (Calvin cycle) and C_4 plants (Hatch and Slack cycle)

C_3 plants (Calvin cycle)	*C_4 plants (Hatch and Slack cycle)*
1. It is found in all photosynthetic plants.	1. It is found only in certain tropical plants.
2. The efficiency of CO_2 absorption at low concentration is far less. So they are less efficient.	2. The efficiency of CO_2 absorption at low concentration is quite high. So they are more efficient plants.
3. The CO_2 acceptor is ribulose-1-5-diphosphate.	3. CO_2 acceptor is phosphor-enol pyruvate.
4. The first stable product is phosphor-glyceic acid.	4. The first stable product is oxaloacetate.
5. All cells participating in photosynthesis have one type of chloroplast (monomorphic type).	5. The chloroplast of parenchymatous bundle sheath is different from that of mesophyll cells (dimorphic type). The chloroplasts in bundle sheath cell are centripetally arranged and lack grana. Leaves shown *kranz type* of anatomy.
6. In each chloroplast, two pigment systems (Photosystems I and II) are present.	6. In the chloroplasts of bundle sheath cells the photosystem II is absent. Therefore, these are dependent on mesophyll chloroplasts for the supply of $NADPH + H^+$.
7. The Calvin cycle enzymes are present in mesophyll chloroplast. Thus, the Calvin cycle occurs.	7. Calvin cycle enzymes are absent in mesophyll chloroplasts. The cycle occurs only in the chloroplasts of bundle sheath cells.
8. The CO_2 compensation point is 50-150 ppm CO_2.	8. The CO_2 compensation point is 0-10 ppm CO_2.
9. Photorespiration is present and easily detectable.	9. Photorespiration is present only to a slight degree or absent.
10. The CO_2 concentration inside the leaf remains high (about 200 ppm).	10. The CO_2 concentration inside the leaf remains low (about 100 ppm).
11. The $^{13}C/^{12}C$ ratio in C-containing compounds remains relatively low (both $^{13}CO_2$ and $^{12}CO_2$ are present in air).	11. The ratio is relatively high, i.e., C_4 plants are more enriched with ^{13}C than C_3 plants.

Contd...

Contd...

C_3 plants (Calvin cycle)	C_4 plants (Hatch and Slack cycle)
12. Net rate of photosynthesis in full sunlight (10,000-12,000 ft.c.) is 15-25 h.mg of CO_2 per dm^2 of leaf area per h.	12. It is 40-80 mg of CO_2 per dm^2 of leaf area per That is, photosynthetic rate is quite high. The plants are efficient.
13. The saturation intensity reaches the range of 1000-4000 ft. c.	13. It is difficult to reach saturation even in full sunlight.
14. Bundle sheath cells are unspecialized.	14. The bundle sheath cells are highly developed with unusual construction of organelles.
15. Only C_3 cycle is found.	15. Both C_3 and C_4 cycles are found.
16. The optimum temperature for the process is 10-25°C.	16. In these plants, it is 30-45°C. So these are warm climate plants. At these temperature, the rate of photosynthesis is double than that of C_3 plants.
17. Oxygen present in air (=21% O_2) markedly inhibits the photosynthetic process as compared to an external atmosphere containing no oxygen.	17. The process of photosynthesis is not inhibited in air as compared to an external atmosphere containing no oxygen. 18. In this process, 30 ATP are required for the synthesis of one glucose molecule.
18. For synthesis of one glucose molecule 18 ATP are required.	18. In this process, 30 ATP are required for the synthesis of one glucose molecule

Difference between cyclic photophosphorylation and non-cyclic photophosphorylation

Cyclic photophosphorylation	Non-cyclic photophosphorylation
1. In the process only PS I is functional.	1. Both PS I and PS II are functional.
2. Electron moves in a closed circle. Electron freed from chlorophyll after excitation to acceptors returns to chlorophyll.	2. In the process water is the ultimate source of electrons and $NADP^+$ is the final acceptor.
3. Reduced NADP ($NADPH_2$) is not formed and assimilation of CO_2 is slowed dawn.	3. $NADPH_2$ is formed which is used in assimilation of CO_2.
4. Oxygen is not evolved.	4. Oxygen as by product is evolved.
5. The system is found dominantly in photosynthetic bacteria.	5. The system is dominant in green plants.
6. The process is not inhibited by DCMU.	6. The process is stopped by use of DCMU.

Difference between respiration and photosynthesis

Respiration	Photosynthesis
1. Oxygen is absorbed in the process.	1. Oxygen is liberated in the process.
2. CO_2 is evolved as a result of oxidation of carbon containing compounds.	2. CO_2 is absorbed and is fixed inside to form carbon containing compounds.

Contd...

Contd...

Respiration	Photosynthesis
3. The process occurs day and night.	3. Process occurs only in presence of light.
4. Light is not essential for the process.	4. Light is essential for the process.
5. During the potential energy is converted into kinetic energy.	5. During the process radiant energy (light energy) is converted into potential energy.
6. Raw materials used are glucose and oxygen.	6. Raw materials used are CO_2 and water.
7. The presence of chlorophyll is not necessary.	7. The presence of chlorophyll is necessary.
8. Energy is released during the process hence it is an exothermic reaction.	8. Energy is stored during the process hence it is an endothermic process.
9. Due to respiration the plant suffers with the loss of weight.	9. By the process, the weight is gained.
10. It is a catabolic process and includes the destruction of stored food.	10. It is an anabolic process and includes the manufacture of food.
11. The process includes dehydrolysis and decarboxylation.	11. It includes the processes like hydrolysis and carboxylation.
12. During the breakdawn of glucose molecule, 38 ATP molecules are formed.	12. During the synthesis of one glucose molecule molecule, 18 molecules are utilized.

Difference between light reaction and dark reaction

Light reaction	Dark reaction
1. Activities found in thylakoids or grana.	1. Activities found in stroma.
2. Photostage reaction.	2. Synthesis stage reaction.
3. Hydrogen transfer phase.	3. Carbon assimilation phase.
4. Temperature quotient is always unity.	4. It is always equal to 2 or 3.
5. The photosynthetic yield rate was higher/faster.	5. It was lesser/slower.
6. The assimilation of CO_2 is slower	6. The assimilation of CO_2 is faster.
7. The reaction is demonstrated by Robert Hill (1937).	7. Three reactions demonstrated by Blackman (1905) and considered Calvin (C_3), Hatch and Slack (C_4) and CAM types.
8. Production of assimilatory powers through PS I and PS II.	8. Utilization of assimilatory powers.
9. The process concerned with cyclic and non-cyclic photophosphorylaton.	9. There was no phosphorylation.

Difference between oxidative photophosphorylation and photophosphorylation

Oxidative photophosphorylation	Photophosphorylation
1. It occurs during respiration.	1. It occurs during photosynthesis.
2. In general it is found inside the mitochondria.	2. It takes place within the chloroplast.
3. The process occurs on the inner membrane of creastae.	3. It occurs in the thylakoid membrane.
4. Molecular oxygen is needed during thermal oxidation.	4. Molecular Oxygen is not required.
5. The energy released during electron transfer due to oxidation-reduction reaction is used during ATP formation.	5. The source of energy for conversion of ATP from ADP and Pi is external (light).
6. The process takes place in electron transport system involving cytochromes.	6. Pigment system P I and P II are involved during the process.
7. The ATP molecules are released in the cytoplasm available for different metabolic reactions.	7. The produced ATP molecules are used up for CO_2 assimilation in the dark reaction of photosynthesis.

Difference between short-day plants and long-day plants

Short-day plants	Long-day plants
1. Flowering is readily induced by exposure to a relatively short day.	1. Flowering is favoured by a relatively long day or even continuous illumination.
2. Short-day plants flower readily under day lengths below the critical, but with day lengths in excess of the critical there is extensive stem elongation without flowering.	2. In long-day plants, on the other hand, there is elongation of the axis followed by flowering with day lengths in excess of the critical, but there is no flowering with day lengths below the critical and most typical plants of this group tend to remain in the leaf rosette stage.
3. For example, *Chenopodium rubrum, Chrysanhemum morifolium, Nicptiana tabacum, Xanthium strumarium, Phgorbitis nil.*	3. For example, *Beta vulgaris, Hordeum vulgare, Hyoscymus niger, Spinacea oleracea.*

Difference between halophytes and glycophytes

Halophytes	Glycophytes
1. Can withstand even 20% of salts in soil and in most cases successfully grow in conditions with 2-6% of salts.	1. Plants that exhibit various degrees of damage and limited growth in the presence sodium salts, usually higher than 0.01%.
2. Potamogetonaceae, Plumbagenaceae, Zygophyllaceae, Frankeniaceae, Tamericaceae, Rhizophoraceae.	2. Conifers, ferns, Orchidaceae, Araceae, Ericaceae and molds.

Difference between drought resistance in plants and cold resistance in plants

Drought resistance in plants	Cold resistance in plants
1. The capacity of plant, to develop normally in dry habitats yielding maximum crop.	1. The capacity of plant, to develop normally under low temperature above the freezing point.
2. Plant capability to endure drought without injury is one of the most important properties of drought resisting plants.	2. Plant capability to endure cold climate without injury is one of the most important properties of cold resisting plants.
3. The categories of plant growing in the areas facing drought are ephemerals, succulents and non-succulent.	3. The underground bulbs, tubers, corms, rhizomes or other such organs that lie below the surface to escape from extreme low temperatures.
4. Drought resisting plants are characterised by higher photosynthetic, respiratory and enzymatic activities; capability of maintaining synthetic reaction during severe wilting period, ability to maintain a high level of hydration, viscosity etc.; capability to endure dehydration and high temperatures and adaptation of morphological features helpful to drought resistance, such as for low transpiration rate.	4. Hardiness by growing under low temp. change in photoperiod; change of unsaturated fatty acids lamellar phase of membrane to globular phase; the quantity of free enzymes, sugar and proteins increases and they become less susceptible to denaturation; free proteins proteins by increasing SH (sulphydrys) content, remove accumulation of inorganic salts by adsorption and sugar countracts the toxicity of electrolytes.
5. Drought resistance can be achieved by using the mechanisms such as rapid phonological development, developmental plasticity, increased rooting, increased hydraulic conductance reduction of water loss, reduction in epidermal conductance, reduction in absorbed radiation, reduction in evaporative surface, solute accumulation, Increased plasticity, protoplasmic resistance.	5. Frost resistance can be induced by water supply, temperature, the kind of concentration of minerals, light intensity and degree of shoot of shoot and root pruning.

Difference between sodicity and salinity

Sodicity	Salinity
1. High concentrations of Na^+	1. High concentrations of total salts
2. Can injure plants directly and degrade soil structure, decreasing porosity and water permeability	2. Ca^{2+}, Mg^{2+}, SO_4^{2-}, NaCl can contribute substantially to salinity.
3. Sodic clay soil (caliche) is so hard and impermeable that dynamite is required to dig through it.	3. Salinity of soil water or irrigation water measured in terms of its electrical conductivity or osmotic potential.

Difference between autoecology and synecology

Autoecology	Synecology
1. It is study of individual species or population in relation to its environment from the time of seed germination upto setting of fruits and seed, i.e., seedling, growth, vegetative growth and reproductive growth.	This is division of ecology deals with the study of plant communities, including their composition, organization and development in relation to environment.
2. This branch deals with study of life cycle, distribution, differentiation, ecological relationships, adaptation and genetic variability of population of individual species in relation to environment.	It is concerned with the structure, nature, development and causes of distribution of communities. his also deals with the distribution of plants on or near surface of earth and water and it also deals with the migration of species.
3. Autoecology forms a basis for the study of synecology.	To understand the ecology of plant communities, the ecology of most important plant species (Autoecology) must first studied.

Difference between predation and parasitism

Predation	Parasitism
1. In this type of association & interaction one species (**predator**) kills & feeds on second species (**prey**).	When two organisms live together in which one derives nourishment at the expenses of the other, the condition is called as parasitism. Parasite usually parasitizes a host, which larger in body size than the host.
2. Predator: animals that hunts & kill the other animals for food.	
3. Prey: animal that is used by other animal as food.	
4. Predator is always larger than the prey.	
5. Predation is an action and reaction in the transfer of energy from one tropic level to other.	The species which provide nourishment is called **Host** and which get support and nourishment is called **Parasite**.
6. The predator kills the prey and used it as a food. It represents a direct and complex interaction between two or more species of eaters and eaten.	Parasite does not kill the host but gets support and nourishment upto completion of its life cycle.

Difference between herbivores and carnivores

Herbivores	Carnivores
1. The animals those are dependent for their food on produces or green plants.	The animals those are dependent on herbivores (i.e. flesh eating animals) for their food.
2. Green plants are the source of energy for herbivores.	Herbivores are in turn becomes source of for flesh eating animals, i.e., carnivores.
3. Herbivores animals are also called as primary consumers and Elton (1939) named herbivores of ecosystem as 'Key industry animals.	Carnivores are also called as secondary consumes consumes, e.g., sparrow, crow, fox, wolves, doges, cats, snakes etc.

Contd...

Contd...

Herbivores	Carnivores
4. Examples: Insects, rodents, rabbit, deer, cow, buffalo, goat etc.	The top carnivores are called 'Tertiary consumers, which prey upon other carnivores, omnivores and herbivores, e.g., lion, tigers, hawk, vulture.

Difference between producers – autotrophs and consumers – heterotrophs

Producers – Autotrophs	Consumers – Heterotrophs
1. Autotrophs are all green plants which fix the solar radiant energy and manufacture food from inorganic substances.	The heterotrophs are non-green plants and all animals which take food from autotrophs.
2. Autotrops are the basic source of energy in ecosystem because they convert the light energy to chemical energy which is locked up in energy rich carbon compounds and oxygen evolved as by product.	The oxygen evolved by plants used in respiration respiration by all living things. The energy in transfers from autotrops to heterotrops through through consumption of food synthesized by producers.
3. Autotrops are the basic tropic levels of ecological pyramid and higher biomass/ energy level as compared to heterotrops.	Heterotrops represents the 2nd, 3rd and upper tropic level of ecological pyramid. The biomass and energy reduces as per the upper tropic level.
4. Algae and other hydrophytes of ponds, grasses of the field, crops, trees, chemosynthetic bacteria, carotenoid bearing purple bacteria are the examples of autotrophs. Parasite, Scanvengers and Sprobes.	Classified in 4 categories. (a) Primary consumers (a) Primary consumers e.g. Herbivores, (b) Secondary (b) Secondary consumers, e.g., Carnivores (c) Territory consumers, e.g., Top carnivores (d) Parasite, Scanvengers and Sprobes.

Difference between heliophyte (sun plants) sciophyte (shade plants)

Heliophyte (Sun plants)	Sciophyte (Shade plants)
1. The stems are thicker and rigid.	The stems are narrow and slender.
2. Internodes are shorter with greater stem branching.	Internodes are long with sparse stem branching.
3. Root system well developed with higher root: shoot ratio.	Roots are fewer and smaller with lower, root: shoot ratio.
4. Leaves are smaller, thick with short petioles.	Leaves are larger, thin with longer petioles.
5. Leaves are arranged vertically with pale green colour with tints of yellow and red.	The flat surface of the leaves face the sun with bright green colour.
6. Leaf surface is often shining of glossy with more hairs per unit area, if leaves are pubescent.	Leaf surface generally dull with few hairs if leaves are pubescent.
7. Leaf cells are smaller with smaller chloroplast.	Leaf cells are larger with large chloroplasts.

Contd...

Contd...

Heliophyte (Sun plants)	Sciophyte (Shade plants)
8. Stomatas are smaller, numerous and restricted to lower surface.	Stomates are larger, less numerous and may occur on both the surfaces.
9. Photosynthetic rate is comparatively less/unit area/unit energy	Photosynthetic rate is higher per unit area/unit energy.
10. Rate of transpiration per unit area is higher.	Rate of transpiration per unit area is lower.
11. Osmotic pressure is higher.	Osmotic pressure is lower.
12. Carbohydrate/nitrogen ratio is higher.	Carbohydrate/nitrogen ratio is lower.
13. There is abundant flowering and fruiting.	There is more vegetative growth with less flowering and fruiting.
14. Light compensation point is higher (100-400 ft. candles)	Light compensation point is lower (2.5-100 ft. candles)
15. Spongy parenchyma is weakly developed.	Spongy parenchyma is well developed.

Difference between food chain and food web

Food chain	Food web
1. A linear sequence of organisms that exist on successive tropic levels within a natural community, through which energy is transferred by feeding.	A non linear network of feeding between organisms that includes many food chains and and hence multiple organisms on each tropic leve.
2. It is the transfer of energy and nutrients through a succession of organisms through repeated process of eating and being eaten.	The complex interrelated food chains makeup the food web.
3. Food chain in any ecosystem runs directly in which green plants are eaten by herbivores, then herbivores are eaten by carnivores and these are eaten by top carnivores.	Food web maintains the stability of the ecosystem. The greater the number of alternative pathways the more stable is the community of living things.

Types of food chain
(1) Grazing food chain
(2) Parasitic food chain
(3) Saprophytic of detritus food chain.

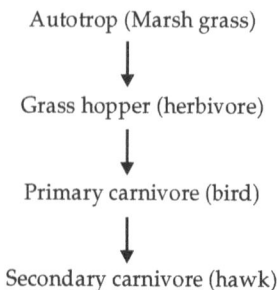

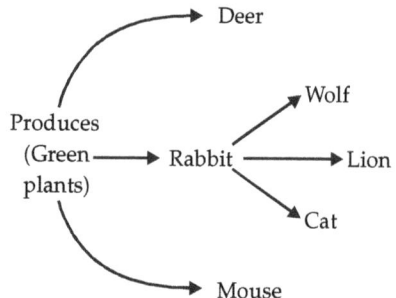

Autotrop (Marsh grass)
↓
Grass hopper (herbivore)
↓
Primary carnivore (bird)
↓
Secondary carnivore (hawk)

Produces (Green plants) → Rabbit → Deer, Wolf, Lion, Cat, Mouse

Difference between xerophytes, hydrophytes and mesophytes

Xerophytes	*Hydrophytes*	*Mesophytes*
1. Plants which grows in dry habitat or xeric conditions *i.e.,* survive under the condition of very poor supply of available water in the habitats, e.g., Astragalus, Asparagus, Ulex, Pinus, Opuntia.	Plants which grow in wet places or in water either partly or wholly submergedare called hydrophytes or aquatic plants, e.g., water Hyacinth, Salvinia, Azolla, Hydrilla and Lotus.	The plants which grow in situations that is neither too wet nor too dry. These plants neither grow in water or water logged soil nor can they survive in dry places.
2. Xerophytes have well developed root system which may profusely branch. Superficial roots are growing deep in the soil.	Root system of hydrophytes are developed poorly developed which may or may not be branched in submerged hydrophytes.	Root system is well developed. These plants of those regions regions where climate and soils are favourable.
3. Stem of xerophytes is very hard and woody, also covered with thick coating of wax and silica. Some stem may be cov-ered dense hairs. Stem is modified into thorns, *e.g.,* Duranta. In stem succulent main stem become bulbous and fleshy and arise from top of roots.	Stem of aquatic plants is very delicate and green or yellow in colour. In some plants, it may be modified into rhizome or runner etc.	Mesophytes classified into (1) Communities of grasses and herbs, (2) Communities of woody plants. Stem is long long. In Deciduous plants, stem thick bark covering on the stem, formation of und-ergroud stem which protects the perennating buds from extreme drought and cold.
4. In xerophytes, if the leaves are caducous, i.e., they fall early in the season, but in majority of plants leaves are generally reduced to scales, e.g., Casuarina, Ruscus, Asparagus. Some evergreen xerophytes have needle-shaped leaves, e.g., Pinus. In leaf succulents, the leaves swell remark-ably and become very fleshy owing to storage of excess amount of water and latex in them.	In floating plants, leaves are generally peltate, long circular, light or dark green in colour, thin and very smooth. Their upper surfaces are exposed in the air but lower ones are generally touch with water. Leaves of free floating hydro-phytes are smooth, shining and frequently coated with wax. The wax coating protects the leaves from chemical and physical injuries and also prevents water clogging of stomata. Leaves of submerged plants are small and harrow.	(1) Communities of grasses and herbs:Includes annual and perennial grasses and herbs. The leaves are thin and glasslands occur in area of approx 25 to 75 cm. rainfall. Stem is slender thin. 2 groups (i) Arctic and alpine alpine mat-grasslands and mat herbage. Plants are small sized, soft shrubs, mosses (ii) Meadow: It is connecting link between mesophytes and hydrophytes, as they grow in in soils where moisture is 60-
5. Leaves with thick cuticle and dense coating of wax/ silica.		83%, e.g., Gramineae, Ranu-nculaceae, and Papilionaceae. (iii) Pasture on cultivated and: vegetation is shorter and open than medow.

Contd...

Contd...

Xerophytes	Hydrophytes	Mesophytes
6. Anatomical modifications: (i) Heavy cutinization, lignifications and was deposition on the surface of epidermis. Shining leaf surface (smooth) of cuticle herbage reflects the rays of light and checks heavy loss of water. (ii) Epidermis cells are thick, small and compact, two or three layered. Cells are radialy elongated, wax, tannin resin; cellulose deposited on surface of epidermis reduces the evaporation of water from plant body. (iii) Hairs on epidermal region or stem/ leaves prevent excess water loss from stomata. (iv) per unit area number of stomata is reduced. Stomata are modified and sunken type reduces transpiration.	Anatomical modification: (i) Reduction in protecting structure: Cuticle is totally absent in the submerged parts of the plant. Epidermis is not protecting layer but absorbs water, minerals, and gases from quatic environment. Hypodermis poorly developed. (ii) Increase in aeration: Stomata are totally absent in submerged parts of the plant. Stomata develop in very limited number and confined only to the upper surface. Exchange of gases takes place directly through cell wall. Air chambers in submerged leaves filled with respiratory gases and moisture. (iii) Reduction in supporting or mechanical tissue.	(2) Communities of woody plants: (i) Mesophytic bushland bushland: These occurs where where temperature of other factor not favourable for the growth of forest Mat herebage vegetation, e.g., Salix, Arabis, Lathyrus, Vicea etc. (ii) Deciduous forests: found in areas where rainfall rainfall is high enough (75-150 cm. per year). These trees trees become leafless for certain period of the year. Phenomenon of repeated foliation and defoliation is prominent in temperate and cold regions (long winter) and tropical regions (long summer). Some plants behave mesophyte in during rainy season and xerophytes during dry cold season are called Tropophytes.
7. Regulation in transpiration: Presence of cuticle, polished surface, compact cells, sunken stomata protected by stomatal retain hairs regulate transpiration. High osmotic pressure of cell sap which increases the turgidity. The turgidity of cell exerts tension force on cell walls, resulted in prevention of wilting of cell and it also affects the absorption of water.	The aquatic plants exhibit a low compensation point and low osmotic concentration of cell sap. Osmotic concentration of cell sap is equal or slightly higher than the water. Nutrients generally absorbed through plant surfaces.	(iii) Evergreen forests: Found in tropical and subtropical regions extending into cold temperate zones of southern hemisphere. There plants retain their leaves for more than one year until new foliage appears. (1) Antarctic forests (2) Subtropical forests (3) Tropical rain forests or tropical evergreen forests.
8. Succulents' xerophytes contain polysaccharides pentosans and a number of acid by virtue of which they are able to resists drought. The stoma opens during night hours and reamin closed during the day.	There is no transpiration from submerged hydrophytes, however, emergent free floating plants have excessive rate of transpiration. Mucilage cells and mucilage canals secrets mucilage to protect the plant body from decay under water.	

Difference between symbiosis and parasitism

Symbiosis	*Parasitism*
1. Symbiosis is a broad term used to describe any close, long-term relationship between species or A relationship between different organisms that live in direct contact.	It denotes a relationship in which one species benefits at the cost of the other. The species which gets benefit is called parasite. The species gets exploited from parasite is called host.
2. Symbiosis types are parasitism, mutualism and commensalium. Symbiosis usually involves supply of food, protection, cleaning, and transportation.	Parasitism is usually for food and support, the association is harmful to host (negative relationship).
3. Positive interaction (+) organism benefits. Negative (–) interaction organism losses. Zero '0' no interaction important in evolution process.	Parasites could have been one of the most powerful forces driving evolution on the plant.

Difference between biotic factors and abiotic factors

Biotic factors	*Abiotic factors*
1. Biotic factors include the influence of living organisms, both plants and animals upon the vegetation. Biotic factors include (i) Interactions between the plants and local animals and man. (ii) Interaction among plants growing in a community (iii) Interaction between plants and soil microorganisms. Biotic components of ecosystem include 2 basic groups: (1) Autotropic components and (2) Heterotropic components.	The physical (abiotic) factors influence the organisms growing on earth includes 3 main media:a) Atmosphere (air) (b) Lithosphere (soil) (c) Hydrosphere (water) A.S. Boughey (1971) classified factors influencing organisms into 4 classes. (i) Climatic factors (related to aerial environment) (ii) Edaphic factor (related to soil conditions) (iii) Physiographic (topographic) factorsiv) Biotic factors.
Autotropic components are all green plants which fix the radiant energy of sun and manufacture food from inorganic substance.	**Climatic factors:** includes (i) Light – radiant energy coming from sun, utilized by autotrops to convert it into food. (ii) Temperatures (iii) Precipitation and atmospheric humidity. (iv) Air and atmosphere.
Heterotropic components include all non-green plants and animals which take food from autotrops. The biotic components are further divided into 3 heads. (i) Producers (Autotropic components) (ii) Consumers and (iii) Decomposers or reducers and transformers.	**Edaphic factors:** includes Earth, i.e., soil, soil type, soil profile, soil reaction, soil nutrients, soil temperature, soil atmosphere, soil organisms. **Physiographic factors:** Topographic features such as elevation and slope on earth, behaviours of earth surface, geo dynamic process.

Difference between diatropism and gradation

Diastropism	Gradation
Uprising of huge earth crust and formation of mountains in parts of continent or whole continent.	Leveling of mountains and high land by geological sinking, erosion and dissolution of rocks.

Difference between lentic environment and lotic environment

Lentic environment	Lotic environment
1. This type of environment found in standing water, i.e., pond, lake etc.	This type of environment found in flowing water, *i.e.,* river, streams.
2. Different plant species are found according to the depth of pond.	Type of vegetation depends upon the speed of water.
3. Light penetration and temperature in deep lake determines the occurrence of different species. Azolla, Hyrilla are producers, Frog, Fishes, Snakes are consumers.	The velocity of current is fast to keep the bottom bottom clear. Filamentous algae and fishes are common producers and consumers respectively.

Difference between tropsphere, stratosphere and mesophere

Troposphere	Stratosphere	Mesosphere
1.The lowest layer of atmosphere in which man and other living organisms live is called Troposphere, is about 20 km above the earth surface.	The second layer of air mass extending about 30 km. above troposphere is called stratosphere	It is third layer of atmosphere next to stratosphere, is about 40 km in height.
2. It is thin in the polar regions (about 10 km thick) and mixture of several gases.	The uppermost layer of stratosphere is called stratopause.	This region is characterized by low atmospheric pressure and low temperature.
3.Troposphere is characterized by steady decrease in temperature and it may decrease height upto –60°C in the upper layer.	In this zone, the temperature shows an increase from a minimum of about –60°C to a maximum of about 5°C.	The temperature begins to drop from stratosphere, goes on decreasing with the increase in the and reaches a minimum of about –95°C at a level some 80 to 90 km abve the earth.
4.The proportion of gases in the atmosphere is fairly constant. Nitrogen 78. 0841%, Oxygen 20.9476% CO_2 0.318%	Ozone is formed from oxygen by a photochemical reaction in which solar energy splits the oxygen molecule to form atomic oxygen which then combines with oxygen molecule to form ozone O_2'! 2O (atomic oxygen)O_2+O O_3 (Ozone)	Ozone cone decreases rapidly with height. Mesosphere separates into from thermosphere.

Contd...

Contd...

Troposphere	Stratosphere	Mesosphere
S. Dust and water vapours are present in troposphere. Troposphere is the layer of sulphates and is the region of strong air movements, cloud formation, lightning, thundering etc.	The absorption of ultraviolet radiation by ozone umbrella is of paramount importance in the ecosystem because these radiations are prevented from it reaching the earth surface where would be lethal to living organisms.	

Difference between primary succession and secondary succession

Primary succession	Secondary succession
Plant succession: The gradu al replacement of one type of plant community by the other is referred to as plant succession.	
1. When the succession starts on the extreme bare area on which there was no previous existence of vegetation, it is called primary succession or presere.	The succession starts on secondary bare area which was once occupied by original vegetation but later became completely cleared of vegetation off vegetation by the process called denudation.
2. This succession starts in the areas of extreme conditions and the process terminates after a long series of intermediate stages. Such bare areas are called as primary bare areas.	This denudation process is brought about by the destructive agencies such as fire, cultivation, strong wind and rains. Such bare area is termed as denuded area or secondary bare area.
3. Primary succession having more intermediate stages and takes more period to reach climax.	Secondary succession has fewer stages and climax is reached very quickly.
4. **Types:** (a) Hydrosere: Sense of changes taken place in aquatic environment (Hydrarch) (b) Halosere: It special type sere which begins on salty soil or in saline water. (c) Xerosere: Succession develops in xericordry habitat. (i) Psammosere-begins on sandy habitat. (ii) Lithosere: Occurs on rock surface. (d) Serule: succession that occurs in microorganisms such as bacteria, fungi, etc.	Secondary succession or subsere develops usually in settled areas greatly modified by the activities of man. This type of successions may develop on secondary bare habitat such as burnt lands, lumbered lands, disturbed agricultural lands etc.

Difference between soil structure and soil texture

Soil structure	Soil texture
1. Arrangement, orientation and organization of soil particles (sand, silt, clay) on certain defined patterns are called soil structure.	The relative percentage of sand, silt and clay (soil particles) in soil is called as soil texture. Soils are classified into different classes on the basis of texture.
2. Soil structure can be changed or altered and influenced by air, moisture, organic matter, altered.	Texture on soil remain unchanged over a long period of time, i.e., can't be changed or microorganisms and root growth.

Soil structure	Soil texture
3. The natural aggregates of soil particles is called as **peds** and artificially formed is called **clod**.	The knowledge of soil texture is of great help in the classification of soil and in determination of degree of weathering of rock.
4. Soil structure reveals the colour, texture and chemical composition of soil aggregates. Types: (1) Prism like (2) Block like (3) Sphere like silt clay etc.	**Types:** Sandy loam, Lomy sand, sandy, loam, silt loam, silt, sandy clay loam, clay loam, clay,

Difference between weathering and pedogenesis

Weathering	Pedogenesis
1. Weathering is disintegration or degradation of complex mineral substances that are locked up in the rocks and eventually simple compounds are formed.	Pedogenesis is soil forming process which leads leads to developed mature soil from the weathering material or parental material or regolith.
2. It disintegration or distractive process in which parent rock broken/fragmented as a result of weathering to form new/altered minerals.	It is constructive process and bio-geochemical nature in which biological influence play important role.
3. It is initial stage of soil forming process.	It is latter stage of soil forming process.

Consolidate parent rock $\xrightarrow[\text{(Weathering)}]{}$ Small particles of parent material of regolith $\xrightarrow[\text{(Pedogenesis)}]{}$ Mature soil with profile differentiation

4. Weathering process consists physical weathering, chemical weathering and biological weathering.	Weathering resulted into development of loose layers or horizons with addition of organic matter (humus), and interaction between minerals and organic matter the horizons formed called soil profile.

Difference between hygrscopic water and capillary water

Hygroscopic water	Capillary water
1. Water which is absorbed on the soil particles and held on the surface of soil particles by forces of attraction and cohesion of its molecules is called hygroscopic water.	The water which is held by surface tension and attraction force of water molecules as thin film around soil particles is capillary spaces is called capillary water.
2. The water retention is > -31 bar	The water retention is $-1/3$ to $-3/1$ bar
3. Non-available water to the plants.	Water available to the plants.

Difference between lichens and lianas

Lichens	Lianas
Lichens are organisms that are a symbiosis between algae and fungus. The photosynthetic algae produce food and fungus provides protection for algae.	Liana is a woody climbing vine that grown on tree trunks in order to reach sunlight in the rain forests.
The morphology, physiology and biochemistry of lichens are very different from those of the isolated fungus and algae in culture.	The woody stems of these plants have well developed alternating vertical columns of secondary xylem and parenchymatous tissue which enables them to twist around supporting objects.
Lichens occur in some of the most extreme environments on earth-arctic tundra, hot deserts, rocky coasts and also abundant epiphytes on leaves and branches in rain forests and temperate woodland, on bare rock.	Lianas tropical moist deciduous forests and rain forests. These climbers form bridges between the forest canopy, connect the entire forest and provide arboreal animals with paths across the forerst.
Lichens must compete with plants for access to sunlight, because of their small size and slow growth, they thrive in places where higher plants have difficultly growing.	The lianas affect other plants also because they cast their shadow and check the light from reaching to the plants of lower storeys.
Lichens are useful to scientists in assessing the effect of air pollution, ozone depletion and metal contamination. For example, collema.	Lianas play an ecological role in providing access routes in the forest canopy for arboreal species. For example, Monkey ladder (*Entada gigas*), Water vine (*Cissus hypoglauca*), Pothos (*Epipremnum aureum*).

Difference between natality and mortality

Natality	Mortality
1. Natality refers to the rate of reproduction unit or birth per unit time expressed in the population by birth, hatching, germination or fission.	Mortality refers to the number of deaths per time.
Birth rate or Natality (B) = $$\frac{\text{Number of birth/unit time}}{\text{Average population}}$$	Mortality rate = D/tD = no. of deaths t = time or period.
2. The maximum number of births produced per individual under ideal conditions of environment is called potential natality.	Potential mortality represents the minimum of theoretical loss of individual under ideal or non-limiting condition.

A birth – death ratio = Births / Deaths x 100
(virtual index)

Difference between emigration, immigration and migration

Emigration	Immigration	Migration
1. It is one way movement of individuals out of population.	It is one way movement of individuals into a population.	It is two way mass movement of the entire population.
2. Emigration occurs when there is over crowding in the population. It leads to reduce the density of unfavourable periods.	It leads to rise in density of population.	It involves the periodic departure and return of the individuals of population and occurs only in mobile organisms during population.
3. Emigration regulates the population on the particular site and prevents over exploitation of the habitat. Also offers individuals a new opportunity to the individuals of population to interbreed leading more heterozygosity.	Immigration may results in decreased mortality among the immigrants or decreased reproductive capacity of the individual.	Migration of population occurs for food, shelter or reproduction. During During migration of population mortality of numerous individuals may occur due to various ecological ecological hazards such as temperature fluctuation, scarcity of food, predation etc. Migration avoids intraspecific competition for food, shelter etc.

Difference between non-degradable pollutants and bio-degradable pollutants

Non-degradable pollutants	Bio-degradable pollutants
A pollutant is substance which may alter environment constituents or cause pollution.	
1. These pollutants are not acted upon by microbes but are oxidized and dissociated automatically.	These pollutants are natural organic compounds which are degraded by biological or microbial action, *e.g.,* domestic sewage.
2. Divided into two classes. (a) Wastes, e.g., glass, plastics, phenolics aluminium cans etc. (b) Poisons, e.g., radioactive substances, pesticides, heavy metals like mercury, lead, cadmium etc.	Consumed and broken down in natural substances, *i.e.,* CO_2 and water by biological organisms.

Difference between renewable and non-renewable resources

Renewable resources	Non-renewable resources
1.The energy generated from natural resources such as sunlight, wind, rain, tides and geothermal heat, which are renewable, *i.e.,* naturally replenished in short period of time.	The natural resources which can not be prodcued produced, re-grown, regenerated or reused on a a scale which can sustain its consumption rate.

Contd...

Contd...

Renewable resources	Non-renewable resources
2.These are infinite resources found in ample quantities in nature and regenerated naturally.	These are finite resources that will eventually dwindle becoming too expensive or too environmentally damaging to retrieve.
3. The worlds largest geothermal power installation is the **Geysers** in **California**.	For example, fossil fuel: **Coal**: Anthracite, bituminous coal and lignite
4. Brazil has one of largest renewable energy gasoline programme in the world involving production of ethanol fuel from sugarcane.	**Petroleum forms :** Heating oil, diesel fuel, **Natural gas :** Methane gas, propane gas **Nuclear energy :** Uranium.
5. Kenya has world's highest household solar ownership rate.	These resources often often exit infixed amount and consumed faster.

Difference between monoecious plants and dioecious plants

Monoecious plants	Dioecious plants
1. Mechanism that facilitate cross pollination. Type of mechanism – dicliny -> staminate and pistillate flowers occur in the same plant, either in same inflorescence – *e.g.,* Castor, Mango, Banana, Coconut or in separate inflorescence, *e.g.,* Maize.Monoecious Plants: Cucurbits, Walnut, Chestnut, Strawberries, Rubber, Grapes and Cassava. In case of monoecious crops like Castor, Cucurbits, some mutants produce only pistillate flowers in the place of both male and female flowers. This feature provide opportunities for use hybrid seed production	Mechanism that facilitate cross pollination. Type Type of mechanism – dicliny -> The male and female flowers are present on different plants i.e., the plants in such species aer either male or female, *e.g.,* Papaya, Date, Plam, Hemp, Asparagus, Spinch. The sex is governed by single gene, *e.g.,* Papaya, Asparagus, hermaphordiet hermaphrodite plants in addition to male or female plants also occurs.

Difference between pedigree method and bulk production method

Pedigree method	Bulk production method
1. Individual plants are selected in F_2 and the subsequent generations and individual plant progenies are grown.	F_2 and subsequent generations are maintained as bulk.
2. Artificial selection, artificial disease epidemics, etc. are an integral port of the method.	Artificial selection, artificial disease epiphytotics etc. may be used to assist natural selection. In certain cases, artificial selection may be essential.
3. Natural selection does not play any role in this method.	Natural selection determines the composition of population at the end of the bulking period.
4. Pedigree records have to be maintained, which is often time consuming and laborious.	No pedigree records are maintained.
5. It is most widely used breeding method.	It has been used only to a limited extent.

Contd...

Contd...

Pedigree method	Bulk production method
6. The segregating generations are space-planted to permit individual plant selection.	The bulk populations are generally planted at commercial planting rates.
7. It generally takes 12 years to develop a new variety and to release it for cultivation.	It takes much longer for development and release of a variety. The bulk population has to be maintained for more than 10 years for natural selection to be effective.
8. It demands close attention from the breeder from F_2 onwards as IPS have to be made and pedigree records have to be maintained.	It is simple, convenient and inexpensive and does not require much attention from breeder during the period of bulking.
9. The size of population is usually smaller than that in the case of bulk method.	Large populations are grown. Natural selection is expected to increase the chances of recovery of Transgressive segregants.

Difference between pure line selection and mass selection

Pure line selection	Mass selection
1. The new variety is a pure line.	The new variety is a mixture of pure lines.
2. The new variety is highly uniform, variation with is pure line variety is purely environmental.	The variety has genetic variation for quantitative characters; it would be relatively uniform in general appearance.
3. The selected plants are subjected to progeny test.	Progeny test is generally not carried out.
4. Pureline variety is the best pure line present in the population. The pure line selection brings greatest improvement in original variety.	The variety is inferior to the best pure line because most of the pure lines included in it will be inferior to the best pure line.
5. Pure line varieties have a narrower adaptation and lower stability.	Wider adaptation and greater stability than a pure line variety.
6. Plants are selected for their desirability. It is not necessary that they have similar phenotype.	The selected plants have similar phenotype since since their seeds are mixed to make a variety.
7. It is more demanding because careful progeny tests and yield trials have to be conducted.	It is less demanding to breeder, if the large number of plants are selected, extensive yield trials are not necessary.
8. Seven to eight years are required to develop new variety.	Six to seven years are required to develop new variety.
9. Selection within pure line will be ineffective unless it has become genetically variable.	Selection with in a variety developed through massselection will be effective since it has genetic variation.
10. The produce of pure line variety is uniform in quality.	The produce is not uniform because variety is mixture of several pure lines, hence differ in grain quality.
11. The variety easily identified in seed certification.	Variety relatively difficult to identify in seed certification.

Difference between self pollinated crops – autogamy and cross pollinated crops – allogamy

Self pollinated crops – Autogamy	Cross pollinated crops – Allogamy
1. Pollen from an anther may fall on the stigma of same flower leading to self-pollination or autogamy.	Pollen grains of flower of one plant are transmitted to the stigmas of flower of another plant, it is known as cross pollination or allogamy.
Mechanisms that promotes self pollination.	
Mechanisms:	**Mechanisms:**
Cleistogamy – Flowers do not open at all e.g. oats	Dicliny or unisexuality – Flowers are either staminate or pistillate. 2 types.
	Monoecy- e.g. Caster, mango, banana, maize
	Dioecy – e.g. papaya, hemp, asparagus, spinach
Chasmogamy – Flower opens only after pollination has been taken place. E.g. wheat, rice, barly Stigmas are closely surrounded by anthers ensures self pollination e.g. tomato, brinjal.	Dichogamy – stames and stigmas in hermaphrodite flower matures at different time, thus facilitating cross pollination-2 types
	Protogyny – e.g. Bajra **Protandry** – e.g. sugarbeets, maize
Flower opens but stigma and anthers hidden by some other floral organs, e.g. Legumes.	Stigmas are covered by waxy film. A stigma does not become receptive until waxy film, broken.
Stigmas became receptive and elongates through staminal columns.	
Geitonogamy – when pollen from a flower of one plant falls on to the stigmas of other flower of same plant. The genetic consequences are same as self pollination e.g. maize.	**Self incompatibility** – failure of pollen from a flower to fertilize stigma of same flower or other flower of same plant.
	Male sterility – absence of functional (viable) pollen grains in hermaphrodite flower
Examples	*Examples*
Wheat, rice, barley, oats, foxtail millet, ragi, pea, gram, groundnut, mung, urd, cowpea, soybean, rajma, lentil, sunhemp, jute, sesamum, linseed, tomato, okra, brinjal, chillies, potato, citrus.	Maize, bajra, rye, brassica, sunflower, castor, niger, sugarcane, sweet clover, cabbage, carrot, cauliflower, onion, cucumber (cucurbitaceae faimily), spinach, garlic, coriander, banana, papaya, strawberry, fig, coconut, grapes.
Self pollination leads to a vary rapid increase in homozygosity.	Cross pollination preserves and promotes heterozygosity in a population.
Populations of self pollinated species are highly homozygous, do not show inhreeding depression, but may exhibit considerable heterosis.	Cross pollinated species are highly heterozygous and show mild to severe inbreeding depression and considerable amount of heterosis.
The aim of breeding method is generally is to develop homozygous varieties.	The breeding method is such as improvement in species without reducing heterozousity to an appreciable degree.
The breeding methods used are, pure line, selection, mass selection, bulk, SSD, pedigree method, back cross method, multi line varieties etc.	The breeding methods commonly used in cross cross pollinated crops may be grouped in two broad categories: (1) Population improvement (2) Hybrid and synthetic varieties. Recurrent selection often exceeds 5% and may reach 30%.
Often cross pollinated species: Cross pollination sorghum, cotton, pigeonpea, safflower.	

Difference between apogamy and apospory

Apogamy	Apospory
1. A form of apomix is development of embryo from synergids or antipodal cells without fertilization. For example, cell of embryo sac in not involved in the development of embryo. This is also termed as somatic parthenogenesis.	Some vegetative cells of the ovule develop into unreduced embryo sacs through series of mitotic divisions and without meiosis. The embryo may develop from egg cell or some other cell of such an embryo sac. For example, Malus, ranunculus, orchids, species of hieraceum.

Difference between protoondy and protogyny

Protoondy	Protogyny
1. Mechanism promoting cross pollination.	Mechanism promoting cross pollination.
2. Type of Dichogamy	**Type of Dichogamy**
Stamens and pistils of hermaphrodite flower may mature at different times there by facilitating cross pollination.	
Stamens mature before pistils, e.g., maize, sugarbeet.	Pistils matures before stamens, e.g., bajra.

Difference between Sterility and Incompatibility

Sterility	Incompatibility
1. Male sterility refers to absence of functional pollen grains in hermaphrodite flower, while female gametes function normally.	Incompatibility refers to the failure of pollen from a flower to fertilize the same flower or other flower on the same plant. First reported by Koelreuter.
2. It occurs in nature sporadically, perhaps due to mutations. It has great value in experimental populations.	It occurs in the nature.
Phenotypic classes of male sterility:	
(1) Structural male sterility (stamens absent, malformed or modified into other floral parts, a lack or malformed microsporogenous tissue or absence of microsporangenesis)	(1) Complementary system of SI : Pistil and pollen together provide substance, which stimulate pollen germination and growth of pollen tube if pollen grain differs in SI genotype from that of the pistil, the germination and growth of pollen having similar genotype is not stimulated, called as stimulatory type of SI
(2) Sporagenous male sterility (apparently normal stamens, abnormal, microsporangenous tissue or PMC/microspore formulation or anomalos meiosis.)	(2) Oppositional system of SI:Pollen and pistil produce such substance, which prevent pollen germination and/or pollen tube growth if the the pollen has the same SI reaction as the pistil. However, germination and growth of pollen tube differing is SI reaction is not inhibited. Most common type of SI sib-divided into (a) Heteromorphic system, (b) Homomorphic system-(i) Gametophytic, (ii) Sporophytic Lewis (1954)

Contd...

Contd...

Sterility	Incompatibility
Types of male sterility: (1) Genetic male sterility : Male sterility resulted due to presence of recessive nuclear gene 'ms' when male sterile plant (msms) crossed with male fertile (MsMs) the F1 (Msms) is male fertile. In F2, 3 male fertile : 1 male sterile ration is obtained	(a) Heteromorphic system flowers of different incompatibility groups are different in morphology. For example, Primila. Distyly condition–Pin flowers–long style short stamens. Thrum flowers–short style, long stamens.
GMS maintained by crossing it with heterozygous male fertile (Msms) plants such mating produces 1:1 male fertile : male sterile plants. Rouging or male fertile plants in necessary to maintain GMS.	Compatible mating possible between pin and thrums flowers, e.g., sweet potato, buck wheat.
Types of GMS (1) Environment insensitive differences (2) Environment sensitive – two types (a) Temperature sensitive GMS (TGMS) e.g., Rice (b) Photoperiod sensitive GMS, e.g., Rice (3) Transgenic GMS – Barnase/Barstar system	(2) Homomorphic system: This system found in majority of SI species, morphological in flowers is not associated with this system. Incompatibility reaction of pollen may be controlled by the genotype of the plant on which it is produced is called sporophytic control or by its own genotypes is called gametophytic control
(4) Cytoplasmic male sterility: This type of male sterility is determined by the cytoplasm. CMS is the result of mutation in the mitochondrial genome (mtDNA). This results in unfavourable nuclear–mitochondrial interaction. The cytoplasm of a zygote comes primarily from egg cell; the progeny of male sterile plants would always be male sterile CMS type male sterility is easily transferred by back crossing with recurrent parent, e.g., ornamental species.	As stated the SI reaction of pollen is determined by its own genotype and not by the genotype of plant on which it is produced, the gametophytic system controlled by single gene 'S', which has so more alleles. (a) Fully incompatible mating, e.g., S1S2 × S1S2 (b) Fully compatible, e.g., S1S2 × S3S4 (c) Partially or half compatible mating, e.g., S1S2 × S1S3, S1S2 × S2S3 Examples : Trifolium, Nicotianina, Lycopersion, Solanum, Peturia, East and Mangelsdorf (1925).
3) **Cytoplasic :** Genetic male sterility This is a case of cytoplasmic male sterility where a nuclear gene for restoring fertility in the male sterile line presents. This is interaction between cytoplasm and nuclear gene.	**Sporophytic control** : The SI reaction of pollen pollen is governed by the genotype of the plant on which the pollen is produced and not by the genotype of pollen, Hughes and Babcock (1950).
'R' gene in heterozygous or homozygous condition restores the male fertility. Thus for transfer of CGMS the nuclear genotype or pollinator strain used must be recessive homozygous (rr), e.g. Sorghum, Maize, Bajra, Rice, Wheat etc.	When incompatible pollen lands on stigma, SI reaction is triggered by two distinct phases (i) Prevention of pollen adhesion, hydration and germination and (ii) failure of pollen tubes to penetrate the papillae. Pollen genotype.

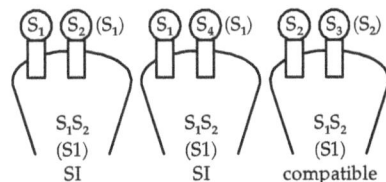

Contd...

Sterility	Incompatibility
	e.g., Radish, Brassica.
(1) Spontaneous mutation : Mutant male sterile cytoplasm arise spontaneously in low frequencies, e.g., maize, bajra and sunflower	Mechanism of SI (1) Pollen – stigma interaction interaction
(2) Interspecific hybridization	(2) Pollen tube – style interaction
(3) Mutations induced through Ethidium Bromide	(3) Pollen tube ovule interaction

Difference between thrum and pin flowers

Thrum flowers	Pin flowers
1. Heteromorphic self Incompatibility.	Flowers of different incompatibility groups are different in morphology.
2. Type of flower in primula	**Type of flower in primula**
Thrum flowers have short styles and long stamens	Pin flowers have long styles and short stamens.

Distyly situation

Compatible mating with pin flowers. This character is governed by single locus 'S' incompatibility of pollen grains determined by genotype of plant on which it produces.	Compatible mating with thrum flowers.

Pin x Pin = ss × ss – Incompatible mating

Thrum x Thrum = Ss × Ss – Incompatible mating

Pin x Thrum = ss × Ss – 1ss (Pin) : 1Ss (Thrum)

Thrum x Pin = Ss × ss – 1Ss (Thrum) : 1ss (Pin)

Difference between pure line and inbreed line

Pure line	Inbreed line
1. Pure lines are occur in species having self pollination	Inbreed lines occurs in cross pollinated crops species.
2. Genetically homozygous in natural populations	Genetic make up of inbreed lines are heterozygous.
3. Obtained through natural self pollination from a single homozygous plant	Obtained through artificial self pollination (or some other form of inbreeding) and selection for several generations.
4. All plants in a single entity are genotipically identical	All plants in a single entity are genotypically almost identical.
5. Used directly as variety	Used for developing hybrid or synthetic varieties.

Difference between qualitative and quantitative characters

Qualitative characters	Quantitative characters
1. Characters are generally governed by one or few genes with large, easily detectable effects, such genes are called as oligonenes, the characters governed by oligogenes are called qualitative characters.	The characters are governed by several genes with small cumulative individual effects are known as polygenes, the characters produced by polygenes are called as quantitative characters.
2. The character produced by oligenes show distinct phenotypic classes. The expression of the oligogenes is uniform.	The character produced by polygenes not shows shows distinct phenotypic classes, and studied by some measurement.
3. These characters are less affected by environment.	The characters are affected by genetic background and environment, and show continuous variation.
4. The phenomenon of single major gene affecting more than one character is known as pleiotropy.	

Difference between autogenous and exogenous variation

Autogenous variation	Exogenous variation
1. Autogenous variation leads to a very rapid increase in homozygosity. Therefore, the populations of such species are highly homozygous.	Exogenous variation promotes heterozygosity in a population. Such cross pollinated species are highly heterozygous.
2. Such self pollinated species do not show inbreeding differences and may exhibit considerable heterosis.	They show mild to severe inbreeding depression and considerable amount of heterosis.
3. The aim of breeding methods generally is to develop homozygous varieties.	The breeding method is such species aim to improve crop without reducing heterozygosity.
4. Selection permits reproduction of only those plants that have the desirable characteristics to tap autogenous variation, with condition that variation must be present in the population and it must be heritable.	The crops showing exogenous variation are highly heterozygous due to the free intermating among their plants. The individual of the population has equal opportunity of mating with any other individual in the population referred as random mating populations.
5. Generally pure line selection, mass selection, pedigree selection approaches are used for crop improvement.	Selection and mating systems are approaches used to improve the crops showing exogenous variation. Mating systems – Random mating, genetic assortative mating, genetic disassortative mating, phenotypic assortative mating, and phenotypic disassortative mating. Hybrid and synthetic varieties are developed in such crops.

Difference between pure line selection and backcross method

Pure line selection	*Back cross method*
1. The new variety is a pure line. The new variety is highly uniform and variation in pure line is purely environmental.	The new variety is identical to recurrent parent parent, i.e., popular variety or pure line with additional one or more characters.
2. The selected plants selfed and subjected to progeny test. The variety is generally the from best pure line present in original population. Pure line selection brings about the greatest improvement over the original variety.	This method is used to transfer the one/more more desirable characters to popular variety from doner parent. Therefore, F₁ and subsequent generations are back crossed to the recurrent parent, i.e., popular variety. The variety developed through back cross method is rectified version of original variety.
3. The new variety has to be extensively tested before release.	Usually extensive testing is not necessary before release.
4. In this method the hybridization between the plants/parents is not followed	Hybridization with the recurrent parent is necessary for producing every back cross generation.
5. The plants are selected for their desirability. It is not necessary that they should have similar phenotype.	The plants should be carefully identified by phenotype for transfer of dominant/recessive gene for back crossing.
6. The plants selected on the basis of desirable phenotype are 200 – 3000.	The back cross generations are small and usually consists of 20-100 plants in each generation.
7. Generally 7-8 years are required to develop the pure line.	For back crossing programme, 12-13 years are required to develop new rectified variety. The breeder should have keep more attention as compared to pure line method.
8. It is not suitable for gene transfer from related species and for producing substitution or addition lines.	It is only useful method for gene transfer from from related species and for producing addition addition and substitution lines.

Difference between single cross and double cross

Single cross	*Double cross*
Hybrid variety is the first generation (F₁) from cross between two pure lines, inbreds, open pollinated varieties, clones or other population that are genetically dissimilar.	
Most commercial hybrid varieties are F₁s from two or more pure lines (rice, tomato) or inbreds (Maize, jowar, bajra).	
Inbred: is a nearly homozygous line obtained through continuous inbreeding of a cross pollinated species, maintained by close inbreeding, i.e., self pollination.	
1. When two inbreds say A and B, are crossed, the hybrid (A x B) is known as single cross.	When two single crosses, say (AxB) and (CxD), are crossed, the resulting hybrid population (AxB) x (CxD), is known as double cross.
2. **Three way cross:** A cross between a single cross (AxB) and an inbred (C) to yield hybrid population (AxB) x C.	Double cross involved 4 inbreds, which are first mated to produced single crosses, the single crosses are then hybridized to yield double cross.

Difference between Recurrent parent and Non-recurrent parent

Recurrent parent	*Non recurrent parent*
1. Recipient parent.	Donor parent.
2. The recurrent/recipient parent is popular variety which lacks in one or two. characteristics, used in back cross method.	The non-recurrent/donor parent is variety/ wild species/land race having one or more specific characters which are lacking in recurrent parent/popular variety, e.g., Rust resistance in wheat, non lodging/short stature in rice, disease/pest resistance.
3. The recurrent parent is repeatedly used in the back cross programme.	The donor parent is used only once in the breeding progrmme for producing the F_1.
4. The sufficient number of back cross (6-7) should be made with recurrent parent so that the genotype of the recurrent parent is recovered in full.	The characters to be transferred from donor parent must be highly heritable and determined by few genes.

Difference between synthetic variety and composite variety

Synthetic variety	*Composite variety*
1. A synthetic variety is produced by crossing in all combinations a number of lines that combine well with each other, i.e., tested for GCA, mixing of equal seed of such all crosses and maintained by open pollination in isolation.	A composite variety is produced by mixing the the seeds of several phenotypically outstanding lines, encouraging open pollination to produce cross in all combinations among the mixed lines lines and maintained by open pollination.
2. The yield of synthetic varieties is predicted in advance because the yield of F_1 obtained from the component lines is available.	The yields of composite variety are not predicted predicted because testing of GCA of lines not available and yield of F_1 are not available.
3. The general combining ability of component lines with each other is tested.	The general combining ability of component lines is not tested.
4. The performance of synthetic variety is better than the composite variety, because former exploit GCA.	The performance of composite variety is usually lower than synthetics.
5. The performance of synthetic varieties is adversely affected by lines with relatively poorer GCA.	Composite variety is mixture of phenotypically outstanding lines.
6. The synthetic varieties are good reservoirs of genetic variability.	The composites are serves as gene reservoirs.

Difference between clone, pureline and inbred

Feature	*Clone*	*Pureline*	*Inbred*
1. Definition	Individuals obtained from a single plant through asexual reproduction (produced by mitotic cell division).	Progeny of a single homozygous self – pollinated plant.	A nearly homozygous line developed by continued inbreeding usually selfing accompanied by selfing.

Contd...

Contd...

Feature	Clone	Pureline	Inbred
2. Mode of pollination in the crop species where they occur.	Cross pollination.	Self pollination.	Cross pollination.
3. Natural mode of reproduction in such species.	Asexual.	Sexual.	Sexual.
4. Genetic makeup of the plants in natural population of such species.	Heterozygous.	Homozygous.	Heterozygous.
5. Maintained through.	Asexual reproduction.	Natural self pollination.	Artificial self-pollination of close inbreeding.
6. All plants in a single entity are genotypically.	Identical	Identical	Almost identical.
7. The genetic make-up of plants within an entity.	Homozygous.	Heterozygous.	Almost homozygous.
8. Used directly as a variety.	Yes	Yes	No (used for developing hybrid or synthetic varieties).
9. Organism where found.	Plant	Plant	Animals, plants.

Difference between population and hybrids

Population	Hybrids
1. Population termed as all such individuals that share the same gene pool, i.e., have an opportunity to intermate with each other and contribute to the next generation of the population.	Hybrids are the first generation (F_1) obtained from cross between two pure lines, inbreds, open pollinated varieties, clones or other populations that are genetically dissimilar.
2. Population of cross pollinated crops is highly heterozygous as well as heterogeneous.	Hybrids are highly heterozygous but homogeneous.
3. Populations are improved through mass selection/recurrent selection with or without progeny test.	Hybrids are developed by crossing 2 parents or using cytoplasmic male sterile line and tested for GCA/SCA and cross seed distributed to farmers.
4. The GCA/SCA testing and top cross was made in recurrent selection.	The GCA/SCA is exploited to full extent.
5. The heterosis not fully utilized in population improvement.	The heterosis is the basis for breeding of hybrids. Heterosis utilized all possible extent.
6. Farmers have not necessary to purchase the seed every year because the inbreeding depression is low.	The hybrids shoe inbreeding depression in F_2 onwards the segregates therefore, farmers have to purchase the fresh hybrid seed every year.

Contd...

Contd...

Population	Hybrids
7. The varieties developed through population improvement, open pollinated, synthetic and composite are less uniform as compare to hybrids.	Hybrid varieties, i.e., produced from hybrid particularly single cross, varieties is more uniform.
8. Population improvement is generally applied in cross pollinated species.	Hybrids can be produced in cross and self pollinated species. It is only possible mean to exploiting heterosis in self pollinated crops as synthetics; open pollinated varieties are not possible in them.
9. For population improvement, it not required much more skill.	For hybrid development requires technical skills, hand emasculation, pollination, utilization and maintenance of CMS lines in isolation. 3 lines (A, B, R) or 2 lines (A and R) in case of TGMS, PGMS are need to maintain separately. In some self pollinated crops, the floral biology (structure) is limiting factor for developing hybrids.
10. The most of these varieties are cheaper than hybrids.	The cost of hybrid seed is relatively higher than the open pollinated varieties.

Difference between vertical resistance and horizontal resistance

Vertical resistance	Horizontal resistance
1. Vertical resistance is generally determined by major genes and characterized by pathotype specificity.	Horizontal resistance is generally controlled by polygenes and is pathotype non-specific.
2. Also called race-specific, pathotype specific or specific resistance.	Also called as race-non specific, pathotype-non specific or partial or general resistance.
3. Host carrying a gene for vertical resistance is attacked by only that pathotype, which is virulent towards that resistant gene, to all other pathotypes, the host will be resistant.	Host carrying a gene for horizontal resistance does not prevent the development of symptoms symptoms of the disease, but slows down the rate of spread of the disease in the population.
4. An avirulent pathotype will produce an immune response, i.e. $r = 0$ or close to zero, but virulent pathotype will lead to the susceptible reaction i.e. $r = 1$.	The reproduction rate of pathogen is never zero, but it is less than 1 and more than 0, i.e. $r > 0$ but $r < 1$.
5. Immune of susceptible response depends on the presence of virulent pathotype when virulent pathotype become frequent, epidemics are common.	Susceptible response not depends or specific race, therefore, the resistance is less than 1 called 'field resistance'.
6. Selection and evaluation is relatively easy.	Selection and evaluation is relatively difficult.
7. Risk of 'Boom and bust' is present (rarely durable).	Risk of 'Boom and bust' is absent (durable)
8. Phenotypic expression is qualitative.	Phenotypic expression is quantitative.
9. Expression-seedlings to maturity.	Express increases as the plant matures.
10. Highly efficient against specific race.	Variable efficiency, but operates against all races.

Difference between physical mutagens and chemical mutagens

Physical mutagens	Chemical mutagens
Agents that induce known as mutagens	Mutations are
Physical mutagens are the different kind of radiations.	Chemical mutagens are the chemical substance used to induce mutations.
(1) Ionizing radiation : (a) Particulate radiation - Densely ionizing a rays, fast neutrons, thermal neutrons sparsely ionizing – b rays (b) Non-particulate radiation – X rays and ã rays both sparsely ionizing. **(2) Non-ionizing radiation :**Ultra violet radiation (UV)	(1) Alkylating agents: Sulpher mustard, nitrogen mustard, epoxide, imines (e.g. ethylene imines (EI), sulphate and sulphonates EMS, methyl-methane sulphonate MMS) diazoalkanes, nitroso compounds e.g. N-methyl–N-nitro-N-nitroso-guanidine or MNNG) (2) Acridine dyes, e.g., acriflavine, proflavine, acridine orange, acrdine yellow, ethidium bromide, (3) Base analogues, e.g., S-bromouracil, S-chlorouracil (4) Other, e.g., nitrous acid, hydroxyl amine, sodium azide.
The primary effect of radiation is ionization. The genetic effect include change in a base, e.g., deamination, loss of a base, breaking of hydrogen bonds in DNA, single and double strand breaks in DNA and cross-linking of DNA strands. Pyrimidines are more sensitive to radiation damage than the pyrines.	The mutagenic changes do not prevent replication, produce changes in one or more nucleotides and do not induce chromosomal aberration. The changes in hydrogen bonding properties of bases or from mistakes in base pairing during DNA replication. Some alterations alterations prevent DNA replication across the altered site, induce chromosome break and chromosome mutations.

Difference between aneuploid and euploid

Aneuploid	Euploid
The changes in chromosome number may involve addition or missing of one or more chromosome from diploid (2x) chromosome complement. In aneuploids, changes are determined in relation to somatic chromosome number (2n) of the species in question. The chromosome number that is not an multiple of the basic chromosome number (x) Types: **Nullisomic**- One chromosome pair missing –2n–2 **Monosomic**-One chromosome missing 2n–1 **Double Monosomic** – One chromosome from each of two different chromosome mussing –2n–1–1 **Trisomic**–One chromosome extra – 2n+1	Individuals whose chromosome number is an exact multiple of the basic number (x) (a) **Autopolyploid** – A polyploidy that has more than two copies on single genome. (b) **Allopolyploid** – A polyploidy containing two or more copies of different non-identical genome. Euploidy is more commonly known as exact polyploidy. Polyploids have larger cell size, large larger pollen grains, seed size. They generally slower in growth and late in flowering.

Contd...

Contd...

Aneuploid	Euploid
Double Trisomics–One chromosome from each of two different chromosome pair extra-2n + 1+1. **Tetrasomics–**One chromosome pair extra-2n+2	Many of our present day crop species, e.g., wheat, N. tabacum, Gossypium spp., Brassica sp. Are polyploids (Euploids) which are commercially cultivated?
Aneuploids are generally weaker than the euploids. Aneuploids seeds are generally smaller, show some what reduced germination and seedlings may show lower viability.	Polyploids are induced by uwing some chemical chemical agents – colchicines, autumnale.
Aneuploids are useful in the production of substitution lines, in the studies on effecr of loss or gain of chromosome or chromosome arm on the phenotype of plant.	
Aneuploids are occur spontaneously in nature or recovered from Autotetraploids, tetrasomic plants etc.	

Difference between autopolyploidy and allopolyploidy

Autopolyploidy	Allopolyploidy
1. A polyploid that have more than two sets of single identical genome	A polyploid that have more than two sets of different/non-identical genome.
Autotriploid – 3 copies of one genome -3x	Allotetraploids- 2 distinct genomes $(2x_1+2x_2)$
Autotetraploid – 4 copies – 4x	Allohexaploid – Three distinct genome $(2x_1+2x_2+2x_3)$
Autopentaploid – 5 copies – 5x	
Autohexaploid – 6 copies – 6x	Allooctaploid – 4 distinct genome $(2x_1 + 2x_2 + 2x_3 + 2x_4)$
2. Autopolyploids occurs spontaneously in the nature, also doubling of chromosome occurs from the adventitious bud leads to callus development at the cut end of stem, and some shoot regenerates from the callus may be polyploid.	Present day allopolyploids were most likely produced by chromosome doubling in F_1 hybrids between two distinct species (distant hybrids) belonging to the same genus or to different genera.
3. Exact doubling of chromosome is possible by giving colchicine treatment to shoot tip meristem (0.2%).	Chromosome doubling might have occurred in somatic tissue due to an irregular mitotic cell division (Amphidiploid).
4. Autopolyploidy has contributed to a limited extent in evolution of plant species. Present day autopolyploid plant species- potato (4x), peanut (4x), coffee (4x), alfa alfa (4x), banana (3x) and sweet potato (6x).	Allopolyploidy have been more successful as many of the crop species of our present day are allopolyploid. Thus allopolyploidy has contributed to a great extent in the evolution of plants, e.g., wheat, cotton, nicotianin spp. etc.
5. Autopolyploidy is more likely to succeed in species with lower chromosome number than in those with higher chromosome number.	Alloplolyploidy is succeeding in species having having lower as well as higher chromosome number.

Contd...

Contd...

Autopolyploidy	Allopolyploidy
6. Cross pollinating species are generally more responsive than self pollinating species.	Many allopolyploids crop species are commercially grown are self/cross pollinating.
7. Crops grown for vegetative parts are more likely to succeed as polyploids than those for seeds.	Allopolyploidy is succeeding in the species where the seed is economic part.
8. Autotetraploids show high sterility accompanied with poor seed set.	The amphidiploids developed through allopolyploidy are highly fertile.
9. Triploids can not be maintained except through clonal propagation.	The maintenance of allopolyploid species is easy through seed production.
10. Allopolyploids are always characterized by a few or more undesirable features, e.g., poor strength of stem in grapes.	Undesirable features are relatively lower in the case of allopolyploid species.

Difference between General combining ability and specific combining ability

General combining ability	Specific combining ability
1. Average performance of a strain in a sense of cross combinations. Estimated from the performance of 'F1's from the crosses.	Deviation in performance of a cross combination from that predicted on the basis of general combining abilities of the parent involved in the cross.

Difference between chilling stress and freezing stress

Chilling stress	Freezing stress
1. When temperature remains above freezing i.e., > 0°C, it is called chilling and resulting stress to plants is termed as chilling stress.	When plants are subjected to subzero temperature temperature (< 0°C), a complex array of stresses and strains develop within them, they are all included in freezing stress.
2. Chilling stress can be measured in terms of seed germination, growth, fruit set, yield, pollen fertility and fruit quality.	As water in plants cools below 0°C it may either freeze, i.e., from ice, or super cool without without forming ice.
3. Chilling stress during germination reduces germination, decline in root growth. It results in the 'locked open' state in stomata. ABA accumulated in chill-affected plants and it may be involved in chilling-hardening.	Initiation of ice formation on plants surface is sufficient to induce freezing of internal water in most plant species. Intra cellular ice formation is accepted as lethal and is due to physical disruption of subcellular structures by ice crystals.
4. Chilling resulting in production of abnormal flowers/fruits and failure to seed and fruit set e.g. sorghum, rice.	Freezing may cause disruptions in the semipermeable properties of plasma membranes. This results in loss of solutes from the cells and it remains plasmolyzed.
5. The ability of a genotype to survive/perform better under chilling stress than other genotypes is called chilling tolerance.	The ability of a genotype to survive freezing stress and to recover and regrow after thawing is known as freezing resistance.

Contd...

Contd...

Chilling stress	Freezing stress
Chilling tolerance involves:	Freezing tolerance involves:
(1) Membrane lipid unsaturation	(1) osmotic adjustment
(2) reduced sensitivity of photosynthesis	(2) bund water
(3) increased chlorophyll accumulation and improved	(3) plasma membrane stability
(4) germination	(4) cell wall, xylem mucilages of the cell inhibit freezing stress
(5) fruit and seed set and	(5) synthesis of cold responsive proteins
(6) pollen fertility	

Difference between drought avoidance and drought esca

Drought avoidance	Drought esca
1. This is the ability of plant to retain a relatively higher level of hydration under conditions of soil or atmospheric water stress.	This is the situation where an otherwise drought drought susceptible variety performs well in a drought environment simply by avoiding the period of drought.
2. The common measure of dehydration avoidance is the tissue water status as expressed by water or turgor potential under condition of water stress.	Early maturity is an important attribute of drought escape and is suitable for environment subjected to late-sown drought stress.
3. Turger pressure can be achieved by reducing transpiration (water saver plants) or increasing water uptake (water spender plants).	Early varieties are having lower leaf area index, lower total evapotranspiration and lower yield potential.

Difference between Alien-addition lines and Alien-substitution lines

Alien-Addition Lines	Alien-Substitution Lines
1. An alien-addition line carries one chromosome pair from a different species in addition to the normal somatic chromosome complement of parent species.	An alien-substitution line has one chromosome pair from a different species in the place of the chromosome pair of the recipient species.
2. Alien addition lines are of little agriculture importance since alien chromosome generally carries many undesirable genes as well.	The alien substitution shows more undesirable effect than alien additions, and as a consequence consequence is of no direct use in agriculture.
3. Viable in polyploid species such as wheat, tobacco, cotton, oat, potato.	Viable/confined to polyploid crop species.

Contd...

Contd...

Ploidy	♀ Recipient species S (4x)	♂ Donor species R (2x)	
Genome	Disease susceptible AA BB	Disease resistant CC	
Gamets	AB	C	
F1 hybrid		ABC (3x)	F1 is usually sterile. Chromosome number is doubled by colchicines. The amphidiploid thus produced is fertile. It is backcrossed to the recipient species S.
	♂ AA BB recipient species S	X ↓ Colchicines AA BB CC (6x) ♀ amphidiploids	
	♂ AA BB species S	X AA BB C (5x) ♀	The progeny is a petaploid with a single C genome from the species R and the full diploid complement of species S.
	♂ AA BB	X AA BB + C ♀ (4x+FEW chromosomes from genome C of species R)	The progeny have full somatic complement of S and a few chromosomes from R. Disease resistant plants are selected and backcrossed to species S.
	AA BB + 1 C and AA BB – 1 + 1C	(4x+1 chromosome from C) (4x – 1 + 1 chromosome from C)	Disease resistant plants similar to species S are selected. Alien-addition monosome has one extra chromosome; substitution monosome has the 2n number of S.
	↓ Selfing ↓		
	AA BB + CC and AA BB – 2 + CC	(4x+1 pair of chromosome from species R) (4x – 1 pair + 1 pair of chromosome from species R)	2n + 2 disease resistant plants are alien-addition lines. Resistant plants with 2n chromosome are alien substitution lines.

Production of alien-addition and alien-substitution lines. A simplified scheme based on two hypothetical species S and R; a chromosome from species R carrying a disease resistance gene is to be added or substituted in the full somatic (2n) chromosome complement of species S, which is susceptible to the disease.

Difference between immunity and susceptibility

Immunity	*Susceptibility*
1. Complete absence of symptoms of a disease even after the host is exposed to the pathogen.	Disease development is profuse and is presumably not checked by the genotype of the host.
2. In immunity, the rate of reproduction of the pathogen is zero, that is r = 0.	In susceptibility, the rate of reproduction of the pathogen is one, i.e., r = 1.
3. Immunity results from prevention of pathogen to reach the appropriate parts of the host.	Susceptibility reaction is classified in relative terms only, i.e. in relation to the reaction of other host varieties and prevailing environment.
4. Immunity resulted due to hypersensitive reaction a group of host cells around the point of infection dies and restricts the establishment of pathogen.	

Difference between drought avoidance and drought tolerance

Drought avoidance	*Drought tolerance*
1. Drought avoidance is the ability of the plant to retain a relatively higher level of 'hydration' under conditions of soil or atmospheric stress.	It is significantly lower level of changes are induced in the genotype than another genotype when both of them are subjected to same level of dehydration.
2. It results in various physiological, biochemical and metabolic processes in plants. The measures of drought avoidance are tissue water status as expressed by water or turgor potential under stress.	When cell loss turgor and dehydrate, the is i) reduce chemical activity of water, ii) increased concentration of solutes and macromolecules, (iii) removal of water of dehydration from macromolecule, (iv) alteration in cellular membranes.
3. This can be achieved by either reducing transpiration (water savers) or increased uptake of water (water spenders).	Measurement of drought tolerance: (1) Maintenance of membrane integrity determined by the leakage of solutes from cells. (2) Plant growth.
Measures of drought avoidance:	Growth under stress is an index drought resistance.·
(1) **Reduced transpiration:** Achieved by closure of stomata in response of water deficit.	• Seedling survival after stress is useful index of dehydration tolerance.·
(2) **Osmotic adjustment:** Osmoregulation is positively associated with yield as allows growth and delayed leaf death.	• Seed germination under osmotic stress created by mannitolor poly-ethylene glycol (PEG) is also measure of drought.
(3) **Abscisic acid (ABA):** Stress hormone plays major role in stress avoidance by effecting stomata closer, reduction in leaf expansion and promotion of root growth.	• The stem reserves are powerful resources for grain filling in stress affected plants during grain filling.
(4) **Cuticular wax:** Avoids transpiration through cuticle.	• Presence of large amount of awns in cereals.
(5) **Leaf characteristics:** Leaf pubescence, altering leaf angle, leaf rolling.	• Proline accumulation appears to be involved in tolerance to water and other stress.

Difference between apomixis and parthenogenesis

Apomixis	Parthenogenesis
1. Seeds are formed but the embryos develop without fertilization. Sexual reproduction is either suppressed or absent.	Embryos develop from the embryo sac without without pollination. Two types: (1) Gonial parthenogenesis – embryo develop from egg cell and (2) somatic parthenogenesis – embryo develops from some cell of the embryo sac other than egg cell.

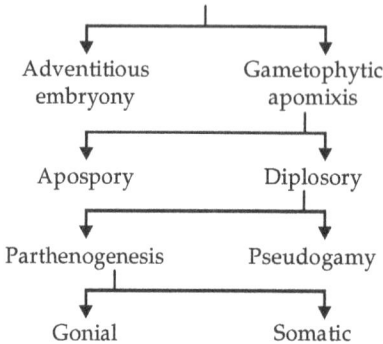

```
                        |
          ┌─────────────┴─────────────┐
          ▼                           ▼
    Adventitious                 Gametophytic
     embryony                      apomixis
          ▼                           ▼
     Apospory                     Diplosory
                                      |
          ▼                           ▼
   Parthenogenesis               Pseudogamy
          |
    ┌─────┴─────┐
    ▼           ▼
  Gonial      Somatic
```

The progeny of apomitic plants is same as the maternal parent.

Difference between certified seed and foundation seed

Certified Seed	Foundation Seed
1. Certified seed is the progeny of foundation and its production is so handled as to maintain specified genetic purity and identity standards as prescribed for the crop being certified.	Foundation seed is the progeny of breeder seed produced by research stations or Government farms or private sector companies.
2. The seed multiplication is certified by State Seed Certification Agency (SSCA).	The seed production is supervised by Seed Certification Agency to maintain the quality according to prescribed standards.
3. Certified seed produced is used for commercial crop production or for truthful seed production.	Foundation seed produced is used for certified seed production.
4. The tag of certified seed (production) is of Blue colour (shade ISI No. 104, azure blue).	The tag of foundation seed is of White Colour.
5. Certified seed production is undertaken on the farms of progressive farmers under strict supervision.	Foundation seed production is undertaken on Government farms, Research Stations and Private Companies.
6. When the seed is progeny of foundation seed, it is called certified seed stage-I, while multiplied seed of stage-I is called certified seed stage-II.	When the seed is progeny of breeder seed, it is called foundation seed stage-I and multiplied seed of stage-I is called foundation seed stage-II.

Both the seeds are certified by a Seed Certification Agency notified under section 8 of Indian Seed Act (1966).

Difference between monocot and dicot

Monocot	Dicot
1. Seed with single cotyledons	Seed with two cotyledons.
(A) **Endospermic/Albuminus:** Fleshy mass around embryo seeds with endosperm, e.g., crops of Gramineae family, sorghum, bajra, wheat, act as food to developing embryo.	(A) **Endospermic/Albuminus:** Seed with endosperm, e.g., castor, cotton, custard apple, papaya.
(B) **Non-endospermic/Ex-albuminus:** Seed without endosperm. *E.g.,* orchids, pathos.	(B) **Non-endospermic/Ex-albuminus:** seed without endosperm, *e.g.,* pulses like gram, pea, frenchbean, and oilseeds such as groundnut, sunflower, and other crops like tamarind, mango, guar and gourds.
2. Hypogeal germination occurs with majority of monocotyledons, e.g., gramineae.	Epigeal germination occurs with majority of dicots except pea, bean gram, and pulses.

Difference between seed and grain

Seed	Grain
1. Seed is used to produce commercial crop, or different stages of seed according to the class of seed.	Grain is produced to use for raw consumption or commercial market sale.
2. Pedigree of seed is known.	Pedigree of grain is not required, it may be the mixture.
3. Seed should be compulsorily certified.	None such condition is applied to grain production.
4. It should satisfy the minimum seed standards.	No such requirement of standards.
5. It should have maximum genetic and physical purity.	Not known, it may be mixture.
6. It should be viable one as per germination standards.	Viability known or unknown.
7. It should be essentially treated with pesticide/fungicide to protect the seed against storage pest and disease.	None treated with any chemicals, since used for consumption purpose.
8. Seed quality is guaranteed.	Grain quality not guaranteed.
9. Seed bag should be sealed, labeled according to class of seed by seed certification agency.	Grains sold as loose and not such certification is essential.

Difference between breeder seed and foundation seed

Breeder seed	Foundation seed
1. Breeder seed is the progeny of nucleus seed and is the source for initial and recurring increase of foundation seed.	Foundation seed is the progeny of breeder seed

Contd...

Contd...

Breeder seed	Foundation seed
2. Breeder seed production is directly controlled by the originating plant breeder who developed the variety or any other institution or qualified breeder recognized by the authorities.	Foundation seed production undertaken by public sector, *i.e.,* research stations, government farms and private sector companies under supervision of seed certification agency.
3. Breeder seed is genetically so pure as to guarantee that subsequent seed class (foundation seed) shall confirm the prescribed standards of genetic purity.	Foundation seed should confirm the all prescribed standards of Seed Certification Agency.
4. Breeder seed tag is Golden Brown in colour.	Foundation seed tag is white in colour.
Seed standards of breeder seed are less stringent than those for nucleus seed, but they are more stringent than those for foundation seed.	

Difference between nucleus seed and breeder seed

Nucleus seed	Breeder seed
1. Nucleus or basic seed is the original or first seed (= propagating material) of a variety available with producing breeder of the crop.	Breeder seed is the progeny of nucleus seed. Breeder seed production is directly controlled by the originating plant breeder.
This seed has 100% genetic and physical purity along with high standards of all other quality parameters.	Breeder seed is source for foundation seed; therefore, it should be genetically so pure and shall confirm the prescribed standards of seed certification.
2. Nucleus seed is multiplied and maintained by selecting individual earhead/pods/ spikes/plants and growing individual plant/pod/spike progenies (plant to progeny or ear to row).	Breeder seed production is undertaken in isolation adopting all packages of practices of crop under strict supervision of concerned breeder or NSC.
3. The process of nucleus seed production repeated every year, therefore, the nucleus seed is available in small quantities.	The breeder seed is produced on large quantity quantity according to the demand/indent of seed and further multiplication.
4. Tag is not used for nucleus seed, but certificate given by the concerned breeder.	Golden brown colour tag is used for seed certification of breeder seed.

Difference between physical purity test and germiantion test

Physical purity test	Germination test
1. Physical purity implies freedom of seed from inert matter and defective seed.	Germination test determines the percentage of seed that produces healthy rood and shoot.
2. The working samples are closely examined, often with the help of magnifying glass to classify the seed into different components, like, pure seed, seed of other variety of same crop, seed of other crop, seed of weeds, inert matter and defective seed.	The seeds are germinated on wet filter papers placed in Petri dishes. The Petri dishes are kept under controlled conditions in an incubator or in culture room, in the temperature 18–22 °C.

Contd...

Contd...

Physical purity test	Germination test
3. Inert matter include sand, straw, stones, pebbles, soil particles, defective seed are broken or shrunken.	The duration of germination test varies from 7-28 days depending upon the crop species.
4. Physical purity = (weight of pure seed / weight of working sample) x 100	Germination % = (total number of seeds germinated/total number of seeds plated) x 100
Working sample = pure seed + seed of other varieties + seed of other crop + weed seed + inert matter	
5. The total amount of permissible contamination by inert matter and defective seed rages from 1% to 5% (carrot). sunflower.	The germination percentage varies according to different crop and class of seed. It ranges from 90% in hybrid maize to 60% in

Difference between seed viability and seed vigour

Seed viability	Seed vigour
The capacity of seed or any plant part e.g., cuttings, to show living properties and growth, i.e., normal seedling under favourable environmental conditions (preferably in the absence of dormancy)	Germination of seed of emergence of seedling may not be as per prediction of seed germination test (*i.e.,* it may low/high). Seed having better germination in laboratory and emergence in field are known as 'high' vigour seeds', whereas poor vigours seeds have poor emergence in field. The sum total of all properties of seed which determines the potential of activity and performance of seed lot during seed germination and seedling emergence in field condition is considered as "seed vigour".
Methods of testing seed viability: (1) Germination test (2) Tetrazolium method, 2, 3, 5 – triphenyl tetrazolium chloride is colourless but develops intense red colour, when it is reduced by ling cells. Seeds are soaked in tap water overnight and are split longitudinally with scalpel so tat portion if the embryo is attached with each half of the seed. One half of each seed is kept in Petri dish and covered with 1% aqueous solution of tetrazolium chloride for 4 hours. The seeds are then washed in tap water and the number of seeds in which the embryo is stained red is determined.	Methods of determination of seed vigour: (1) **Seed size:** Seeds with high seed weight are vigourous. (2) **Seed density:** Seed lot with high density i.e, bold seed is considered as vigourous. (3) **Physical soundness:** Undeveloped, undersize, shriveled, discoloured and damaged seeds are poor in vigour. (4) **Fresh seeds** with proper dormancy are more vigours than old. (5) High germination at first count indicates high vigour. (6) **Speed germination:** Number of normal seedlings in germination test is counted on every day from 1st day up to final count index of speed.
Viable seeds (%) = (number of half seeds stained red / total number of half seeds) x 100	Germination = Σ(no. of normal seedlings germinated/no of days on which it is germinated

Seed viability	Seed vigour
(3) Embryo excision (EE) method: The test is suitable for sloe germinating seeds (after cutting hard coat) are soaked in water for 4 days at 15°C or low temperature. Embryo of such seeds are excised with sterile scalpel and needle and then placed on moist filter paper in Petri dish at 20°C. Germination of embryo examined after 14 days for differentiation and growth.	(7) **Seedlings growth rate:** The length of each seedling is measured in seed lot after prescribed germination period and average length is worked out. The lot showing maximum seedlings length is considered as vigours. (8) **Seedlings dryweight:** The lot exhibiting maximum dry weight is considered as vigours.
(4) Accelerated Ageing (AA): Seeds are aged at 100% relative humidity and at high temperature (40-45°C) for periods ranging up to 7 days before germination. The difference in standard germination % and percentage after accelerated ageing test indicates the physiological capability of seed to withstand stress during storage.	(9) Mobilization efficiency: Seeds with higher mobilization efficiency are called vigourous having capacity to supply more food to seedlings. 10) Accelerated aging test (11) **Brick gravel test:** This test conducted to identify pathogenic infection in seed. Seed are placed in brick granules. As seedlings emerge, the brick granules placed stress having high vigour.
Other tests to determine seed viability: (1) Indigo carmine (IC) test.	(12) **Tetrazolium test:** Application of seed vigour test.
(2) Radiographic (X ray contrast method) (3) Glumatic acid decarboxylase activity	(1) It indicates which seed lot is of high or low low vigour.
(4) Seed leachate conductivity test (5) Seed crushing test.	(2) It indicates which seed lot performs better under stress sowing conditions. (3) Suitability of seed lot for storage. (4) With standing seed lot during transport.

Difference between normal seedling and abnormal seedling

Normal seedling	Abnormal seedling
1. It is defined as seedling which shows capacity for continuous development into normal plant when grown under favourable conditions for germination.	The seedling which do not have capacity to develop into normal plant when grown in favourable conditions for germination.
2. The seedlings are well developed, complete, proportionate and healthy essential structures such as (i) Long and slender roots (ii) Well developed shoot axis with epicotyl/ hypocotyls/ mesocotyl as per crops. (iii) Specific number of cotyledons cone in monocots and two in dicots. (iv) Green primary leaves: one in monocot and two in dicots.	In such seedlings, one or more essential structures of seedling are defective. (i) Primary root stunted, retarded, broken, splitted, glassy, and trapped in seed coat negative geotropism. (ii) Hypocotyls/mesocotyls/lepicotyls tightly twisted, looped structures. (iii) Cotyledons are deformed, swollen, broken, missing, discoloured, necrotic, glassy and decayed due to primary infection and size ½ or more than half.

Contd...

Contd...

Normal seedling	Abnormal seedling
(v) When more than one seedling is produced by seed, that seed is known as mutigerm unit. E.g., sugar beet. **Normal seedlings %** = $(N_1 / T) \times 100$ N_1 = No. of normal seedlings in 4 replications. N_2 = No. of abnormal seedlings in 4 replications. T = Total no. of seed kept for germination. **Abnormal Seedlings %** = $(N_2 / T) \times 100$	(iv) Primary leaves deformed, missing, damaged, necrotic (dead) due to primary infection and size less than $1/4^{th}$ of normal size. (v) Terminal bud deformed damaged, missing, decayed due to primary infection. (vi) Seedling as whole is deformed, fractured, white coloured, spindly, glassy or decayed due to primary infection.Classes of abnormal seedlings: (i) Damaged seedlings (ii) Deformed seedlings (iii) Decayed seedlings

Difference between viability test and grow-out test

Viability test	Grow-out test
1. Seed viability is determined as per cent of seed that produce or are likely to produce seedlings under a suitable environment. Seed viability is test conducted in laboratory. The tests are (1) Germination test, (2) Tetrazolium text.	Test of genetic purity (genuineness) of the given seed sample in the field on the basis of morphological characters of the plants produced by seed of seed sample under test is termed as 'grow-out test'.
2. In this test, the genetic purity is not tested only capacity of seed to produce normal seedlings is tested.	In this test, authentic sample is planted after every 10 test samples for close comparison observations are made both on qualitative and quantitative traits of the test and the authentic sample plot during entire growing period.
3. This test is suitable for self as well as cross pollinated crops.	The frequency of off types is tested as compared to authentic sample in field.
4. This quick test required less time, efforts and funds.	Grow out tests are more useful in self pollinated than in cross pollinated crops. It requires a much longer time, an excellent green house/off-season nursery facility, much greater efforts and funds.
5. This test is widely used for testing viability in STL.	This test is rarely used because it requires land, time more manpower, funds etc.

Difference between breeder seed and certified seed

Breeder seed	Certified seed
1. It is progeny of nucleus seed or breeder seed.	It is progeny from foundation or certified seed.
2. Produced by original or sponsored plant breeder.	Produced by State Seed Corporation.
3. Produced at experimental farms of research institutes or Agril. Universities.	Produced in the fields of progressive farmers.

Contd...

Contd...

Breeder seed	Certified seed
4. Genetic purity is cent percent.	Genetic purity is 99.9%.
5. Physical purity is 100%	Physical purity is 98%.
6. Certification not required. Inspected by monitoring team consisting original breeder with one representative of NSC and SSCA.	Certification is done by SSCA.
Used to produce foundation seed.	Distributed to farmers for commercial cultivation.

Difference between submitted, working, composite & official samples

Submitted samples	Working samples	Composite samples	Official samples
1. Required quantity of seed sample taken from composite sample for different seed test is known as 'submitted sample'	Small portion of specified weight of submitted sample taken for specific seed test in STL by concerned officer is called working sample.	Primary seed samples drawn from different seed containers/bulks are combined and then mixed thoroughly so as to form/ constitute 'composite seed sample'.	It is submitted and drawn by quality control officer (QCO) or seed low enforced officer from sealed containers of certified seed/fertilizer/ fungicide/insecticide/ pesticide to confirm their quality as per labels or information given on containers.
2. Before taking this sample, composite seed sample should be thoroughly mixed and reduced up to prescribed weight with help of seed dividers or by repeated halving methods.	It is prepared by using dividers or by hand methods.	Its size is 10 times more than required submitted samples.	
3. The lot no., name of crop, date of harvesting, variety, class of seed, date of sampling etc. should be mentioned on submitted sample.	The working sample kept n paper bag marked with code number, hname of crop and purpose.		
Types: (submitted to) (1) Service sample– Private agency (2) Certified sample–SCO (3) Official sample– drawing officer	The weight of working sample is usually 25 gm.		

Difference between pollen shedder and volunteer plants

Pollen shedder	Volunteer plants
1. In hybrid seed production programme using CMS lines, the removal of plant showing viable pollens on anthesis in male sterile line is called pollen shedder.	The plant that germinated from the seeds of earlier crop or accidentally planted in the seed production plot is called as volunteer plants.
2. In maintenance programme, the plant shedding viable pollen in 'A' line should be removed immediately to avoid further cross pollination/contamination.	To retain physical purity of seed roughing of volunteer plant before flowering is essential.

Difference between seed dormancy and seed viability

Seed dormancy	Seed viability
1. Failure of germination of fully developed, mature and viable seed under optimum environmental conditions for germination due to dormancy period is known as seed dormancy.	Capacity of seed or any pant part (cuttings) to show living properties like germination and growth, i.e., normal seedling under favourable environmental conditions. (Preferably in the absence of dormancy.)
2. Seed dormancy due to following factors: (1) Impermeability of seed coats to moisture (hard seed) (2) Impermeability of seed coat to oxygen. (3) Embryo dormancy (4) Presence of inhibitors (5) Light requirement (6) Mechanical restrictions (7) Unfavourable environmental conditions.	Methods of testing seed viability: (1) Germination test (2) Tetrazolium method (3) Embryo Excision (EE) method. (4) Accelerated Aging method.

Difference between variety, hybrid and composites

Variety	Hybrid	Composites
1.It is brad term, in plant breeding, a strain released for commercial cultivation by variety release committee.	Progeny obtained from the crossing of two or more genetically diverse parents (strains). The cross seed between two diverse strains is used for commercial cultivation.	Varieties produced by open-pollination among a number of outstanding strains usually not tested for combining ability with each other.
2.Variety consists of cultivar developed by all breeding approaches in self and cross pollinated species including hybridization.	The hybrid varieties developed by selecting proper parents, selfing, emasculation/use of CMS line, pollination, selfing and harvesting of F_0 seed. Selfing of F_0 seed rise to F_1 commercial crop.	This is population improvement approach. Variety is produced by mixing the seeds of several phenotypically outstanding lines and encouraging open pollination to produce crosses in all combination among the mixed lines.

Variety	Hybrid	Composites
3. Farmers have not to purchase fresh seed every year depending upon the type of variety.	The lines used to produce hybrids are tested for GCA and SCA Farmers have to purchase fresh seed every year.	The lines used to produce composite are not tested for GCA.The seed of composites may be utilized for 3–4 years.

Difference between normal seedlings and abnormal seedlings

Normal seedlings	Abnormal seedlings
1. The seedlings which show the capacity for continued development into normal plants when grown in good quality of soil and under favourable conditions of water, temperature and light.	The seedlings which do not show the capacity of continued development into normal plants when grown in good quality of soil and under favourable conditions of water, temperature and light.
2. The seedlings with all their essential structures well developed or slightly defect of their essential structures. For example, intact- seedlings, seedlings with slight defect.	The seedlings with any of the essential sturcture missing, weak developed, diseased or decayed prevent to normal development. For example, damaged seedlings, reformed or unbalanced seedlings and delayed seedlings.

Difference between certified seed and foundation seed

Certified seed	Foundation seed
1. Certified seed is the seed which is certified by a Seed Certification Agency under section 8 of the Indian Seed Act (1966) or it is the progeny of foundation, register or certified seed.	Foundation seed is a progeny of breeder seed.
2. Its purity is certified by a Seed Certifying Agency.	Its purity is certified by a SCA.
3. It is used for commercial cultivation on farmers' field. It is source of truthful seed. Generally it is for commercial crop production.	It is the source of register and certified seed.
4. Certification was issued once for certified seed not eligible for further seed increase under certification.	Used for further seed increase.
5. Certified seed having a blue colour lable tag.	Foundation seed having with white colour lable tag.
6. Genetic purity of certified seed is 99.00% 99.5%. (var, synthetic, composite, multiply 98.00%).	Genetic purity of foundation seed is

Difference between breeder seed and foundation seed

Breeder seed	Foundation seed
1. Breeder seed is the seed or vegetative production material produced by the breeder who developed the particular variety.	Foundation seed is the progeny of breeder seed.
2. It is the source of foundation seed.	If the source of Register or certified seed.
3. Breeder seed is genetically pure seed (99.9%)	Foundation seed is genetically pure seed (99.00%)
4. Breeders' seed is having with golden yellow colour lable tag.	Foundation seed having with white colour lable.
5. Genetic purity is 99.9% or 100%.	Genetic purity is 99.5%.

Difference between nucleus seed and breeder seed

Nucleus seed	Breeder seed
1. Nucleus seed is a group of progeny of individual plants taken at random from a variety for the purpose of purifying that variety of mixture and that off types and produced by concerned breeder.	Breeder seed is the progeny of nucleus seed in a certification programme directly controlled by the originating or sponsoring plant breedings institute.
2. Nucleus seed have 100% genetic purity.	Breeder seed having maximum genetic purity (99.9% or 100%)
3. Nucleus seed is original source of all classes of certified seed.	Breeder seed is source of foundation seed.

Difference between physical purity test and germination test

Physical purity test	Germination test
1. The determination of different components of the variety viz., pure seed, other crop seed, weed seed and inert matter.	The emergence and development of seedling from seed embryo those essential structures which for the kind of seed being tested.
2. The objective of physical purity test is to determine the submitted sample conforms to the prescribed physical quality standards with regards to purity components.	The objective of the germination test is to determine the percentage of germination to prescribed standards for that particular variety.
3. This test includes separation of different components of seed.	In germination test different crops are tested on different media, e.g., TP, BP, sand.
4. This test is regarding to seed genuineness for seed quality.	The test is regards to the viability of seed of particular crop for particular period.

Difference between seed viability and seed vigour

Seed viability	Seed vigour
1. Seed viability is defined as ability of seed to live, grow and develop under favourable environmental conditions.	Vigour is the sum of all total attributes, which favours rapid and uniform stand establishment in the field.

Contd...

Contd...

Seed viability	Seed vigour
2. A viable seed are which is capable of germinating under the proper circumstances.	Vigour of the seed may also depend on different environment and soil factor, i.e., abiotic and biotic factor.
3. Such viable seed may or may not be redially of immediately germinate.	Old seed vigour is less as compared to fresh seed.
4. Dormant – viable seed may require lengthy specific treatment before they become immediately germinable.	Different vigour test are used to determination of vigour physical, physiological and biochemical.

Difference between monocot and dicot

Monocot	Dicot
1. Synonym of monocotyledon.	Synonym of dicotyledon.
2. It refers to plant that have single seed leaves.	It refers to plant which have two seed leaves in the seed.
3. These plants have parallel vein in their leaves.	The veins are branched in their leaves.
4. No distinct bark and wood layer	Dicot stem always have definite wood and bark layer.
For example, cereal crops.	For example, pulses and oilseeds crops.

Difference between submitted sample and working sample

Submitted sample	Working sample
1. Submitted sample is the sample submitted to a seed testing laboratory.	Working sample is a reduced sample taken from the submitted sample in the laboratory on which one of the seed quality test is made.
2. From composite sample, smaller samples obtained (i.e., submitted sample) by mixing composite sample followed either by progressive subdivisions or by the abstraction and combinations of small portions at random.	Working sample is a sample taken for particular test from the submitted sample.
3. The sample is submitted by the seed certification officer.	The sample taken by seed analyst for testing.
4. Submitted sample should be 1 kg for maize, barely, wheat, 900 gm for sorghum, 400 gm for paddy.	Working sample depends on the tests carried out e.g. for purity 120 g wheat, barely, 900 gm for maize, 90 gm for sorghum.

Difference between viability test and grow-out test

Viability test	Grow-out test
1. The viability test performed to determine the ability to live, grow and develop the seed.	Grow-out test performed to determine the genuineness of seed as to species or variety or freedom from seed born infection.
2. The test performs to test the seed for ability of seed to germinate.	This test performs to seed purity, genuineness.

Contd...

Contd...

Viability test	Grow-out test
3. This test conducted at laboratory	This test conducted at on field.
4. This test is not compulsory to seed certification.	This test is compulsory for to test the purity of different class of seed.

Difference between seed and grain

Seed	Grain
1. Seed is a mature ovule consisting of an embryonic plant together with a store of food, all surrounded by a protective coat.	Grain is a mature or immature ovule consisting of store of food all surrounded by a protective coat.
2. It is generally used for production of crop.	It is generally used as food for human and animal, birds.
3. Seed should be always viable.	It should not need to be viable.
4. The embryo should be live.	Embryo need not to be live.
5. Seed should required proper storage condition to maintain the seed quality.	Grain has not required maintain as like seed.
6. Endosperm supplying nourishment to the embryo during seed germination.	Endosperm need not to supply nourishment to embryo.

Difference between composite sample and working sample

Composite sample	Working sample
1. Composite sample is the sample formed by combining all primary samples obtained from taking small portion at random from different portion in the lot.	Working sample is a reduced sample taken from the submitted sample in the laboratory on which one of the seed quality test made.
2. Composite sample obtained from primary sample.	Working sample obtained from submitted sample.
3. From composite sample, mixing and progressive subdivision or by abstraction submitted sample are formed.	Small portion of submitted sample required for particular test is a working sample.
4. It is from seed lot at seed storage.	It is taken in the laboratory from submitted sample.

Difference between official and working sample

Official sample	Working sample
1. Official sample is a sample received from the seed inspector.	It is the reduced sample with required weight obtained from the submitted sample.
2. It is the submitted sample to seed analyst for seed analysis.	It is the sample drawn from submitted sample for particular test.
3. The use of sample for submission to seed test laboratory.	The use of sample for to test the quality test conducted in seed testing laboratory.

Weight of samples

Crop	Size of seed lot	Submitted sample	Working sample
Paddy	20,000 kg	400 gm	40 gm
Sorghum	10,000 kg	1000 gm	90 gm
Wheat	20,000 kg	1000 gm	120 gm
Maize	40,000 kg	1000 gm	900 gm

Difference between hybrid and variety

Hybrid	Variety
1. Hybrid means a progeny from hybridization between two or more strains and hybrid variety is the F_1 generation from a cross between two different varieties or other population.	A strain released for commercial cultivation by a variety release committee.
2. Hybrid developed form hybridization between two or more strains of desired characters.	Variety developed from the selection of desired character genotypes.
3. Hybrid seed needs to replace every year.	Not need to replace 3 years or more.
4. Care should be taken for maintain the genetic purity.	It is easy to maintain genetic pyrity.

Difference between composites and synthetics

Composites	Synthetics
1. Varieties produced by open pollination among a number of outstanding strains usually not tested for combining ability with each other.	Synthetic variety is cross pollinated species; a variety obtained by making in all possible combinations a number of lines.
2. Composite variety is maintained by open pollination in isolation.	Synthetic variety is maintained by open pollination in isolation.
3. The yield of composite varieties cannot be predicted in advance for the obvious reason that the yields of all the F_1s among the component lines are not available.	The yield of synthetic variety can be predicted in advance.

Difference between composites and variety

Composites	Variety
1. Varieties produced by open pollination among a number of outstanding strains usually not tested for combining ability with each other.	In plant breeding, a strain released for commercial cultivation by a variety release committee. In botany, a subdivision of a species based on from function or both.
2. All the individual plants within composites are not identical.	The entire plants individual within a variety have identical genotype.
3. The yield can not be predicted in advance.	The yield can be predicted in advance.

Difference between seed viability and seed dormancy

Seed viability	Seed dormancy
1. Seed viability is defined as ability of seed to live, grow and develop under favourable environmental conditions.	Seed dormancy is an internal condition or stage of a viable seed that prevents its germination although good growing temperature and moisture provided.
2. Seed viability loss during the ageing (time) hence proper storage conditions are required to maintain viability.	Dormancy is a resting state that must be broken by time or special condition before the seed will germinate.
3. If viability losses, seed cannot germinate.	After dormancy break, seed can be erminated.

Difference between pollen shedder and volunteer plants

Pollen shedder	Volunteer plants
1. In hybrid seed production involving male sterility. The plats of 'B' line present in 'A' line are termed as pollen shedders.	Volunteer plant is an unknown plants growing from seed that remains in the field from a previous crop.
2. Pollen shedder removes before flowering to maintain the hybrid.	Volunteer plants are off type plants in seed production plot and rouging should be done as identified and must before the flowering to avoid admixture of genetic purity of seed plot.

Difference between endospermic and non-endospermic

Endospermic	Non-endospermic
1. Endospermic seeds are those dicotyledons where endosperm is present, e.g., castor, papaya, etc.	Non-endospermic seed are those dicotyledons where endosperm is absent, e.g., gram, pea, bean, sunflower.
2. It is also called as Albuminous	It is also called as Ex-albuminous
3. Albuminous monocotyledonous seed which contain endosperm, e.g., wheat, rice, maize, onion.	Ex-albuminous seed which are monocotyledonous with the endosperm, e.g., water plantain, orchid
4. Monocotyledonous seed are mostly albuminous.	While among dicotyledonous both are common.
5. The food material is stored in the endosperm and hence cotyledons are thin and small	Cotyledons store up of food and become thicky and fleshy.

Difference between seed certification agency and seed law enforcement agency

Seed Certification Agency	*Seed Law Enforcement Agency*
1. The objective of the seed certification agency is to maintain and make available seed to the public through certification.	The object of the seed law is for regulating the quality of certain notified kind/varieties of seed for sale and for matters connected therewith.
2. Seed certification agency formed under the seed law act 1966.	The seed law passed in 1966 at parliament and enforced during the October 1969.
3. Seed of any those varieties which are notified under section 5 of the seed act shall be eligible for certification.	The seed law act extent to the whole of India and it has 25 sections.
4. Seed certification agencies have all the responsibilities of seed certification as notified and kind of variety.	

8

CONCEPTS, THEORIES AND LAWS

Concepts of Heredity

Mendel's Principles of Heredity

Mendel chose two plants of *Pisum sativum* L. differing in a pair of contrasting characters, e.g., a plant with round seed-coat and another with wrinkled seed-coat, as parents for each of his experiments. He than confirmed that they bred true for several generation on self-fertilization.

He then crossed them and obtained the hybrid seeds. He found that the first generation (i.e., the first filial, or F_1 generation) hybrids were always uniform. He than self-fertilized the F_1 and obtained as large a number as possible of a second generation, or F_2. He found that the F_2 consisted of different kinds. He classified the F_2 according to the characters exhibited and counted the number of each class.

In the light of present knowledge, Mendel's principles of heredity can be expressed as follows:

In sexual reproduction, the individual (or zygote) is formed by the fusion of two gametes, one (the egg) from the mother and the other (the sperm) from the father. The hereditary particles are called genes (or factors). The female gamete contributes one of gene from the mother and the male gamete, one of each kind of gene from the father. A zygote carries therefore every gene in duplicate. These genes however do not blend but preserve their individualities.

When this individual forms its own gametes, the maternal and paternal members of each pair of genes segregate and pass to different gametes.

Each gamete therefore has only one member of a pair genes existing in adult individuals.

The members of different maternal and paternal pairs of genes segregate independently and different gametes produced by the same individual may therefore contain different sets of genes.

These principles were summed up by Carl Correns, one of the rediscoveries of Mendel, in what are now known as Mendel's laws of heredity.

The first law is that hereditary factors (genes are found in pairs in mature individuals. They do not blend but separate or segregate unchanged during the formation of gametes. The gametes therefore contain only one of a pair of factors responsible for each character. Even hybrids therefore produce gametes, which are 'pure'.

The second law is that the members of different pairs of factors responsible for different characters segregate and recombine independently in different gametes.

Mendel's Work

From the original research paper of Mendel it was obvious that Mendel was well acquainted with the scientific literature related to hybridization of his time. His approach was simple, logical, scientific, mathematical and analytical. Mendel concentrated his attention on a particular character and at a time, he studied only one character of the hybrid. Further, unlike other hybridists, he designed his hybridization experiments to record the number of different types of the progeny.

Mendel's Selection of the Experimental Plant

He found the plants of family Leguminoseae such as peas and beans, most suitable materials for his experiments, because these plants (1) were easy to culture in open ground or in pots; (2) had short growth period and life-cycle; (3) had self-pollinating flowers of peculiar structure; (4) had contrasting heritable characters, and (5) might produce fertile hybrids on artificial cross pollination. Thus, Mendel found edible pea (*Pisum sativum*) a best material for his hybridization experiments, because its various available varieties (about 34) showed clear-cut differences.

According to the number of pairs of contrasting characters in the parental generation, the resultant hybrids are called as follows.

Monohybrid: Have one pair of different character.

Dihybrid: Have two pairs of different characters, or

Polyhybrid: Have more than two pairs of different characters.

Those characters which were transmitted unchanged and expressed in the hybrid in the hybridization process were called as dominant characters and those which became latent in the process were called as recessive characters by Mendel.

Mendel's Laws

Mendel himself did not postulate any genetical principle or laws as erroneously described various textbooks of genetics. He simply gave conclusive theoretical and statistical explanations for his hybridization experiments in his research paper. It was Correns the discoverer of Mendel's work who thought that Mendel's discovery could be represented by the two laws of heredity. These laws of heredity are "Law of segregation and the Law of independent assortment" or "Law of free recombination." The phenomenon of dominance has been considered erroneously as the law of dominance in some of the textbooks of genetics.

Mendel's Law of Segregation

Mendel's first Law of inheritance, the Law of segregation or **Law of purity of gametes** states that in a heterozygote a dominant and a recessive allele

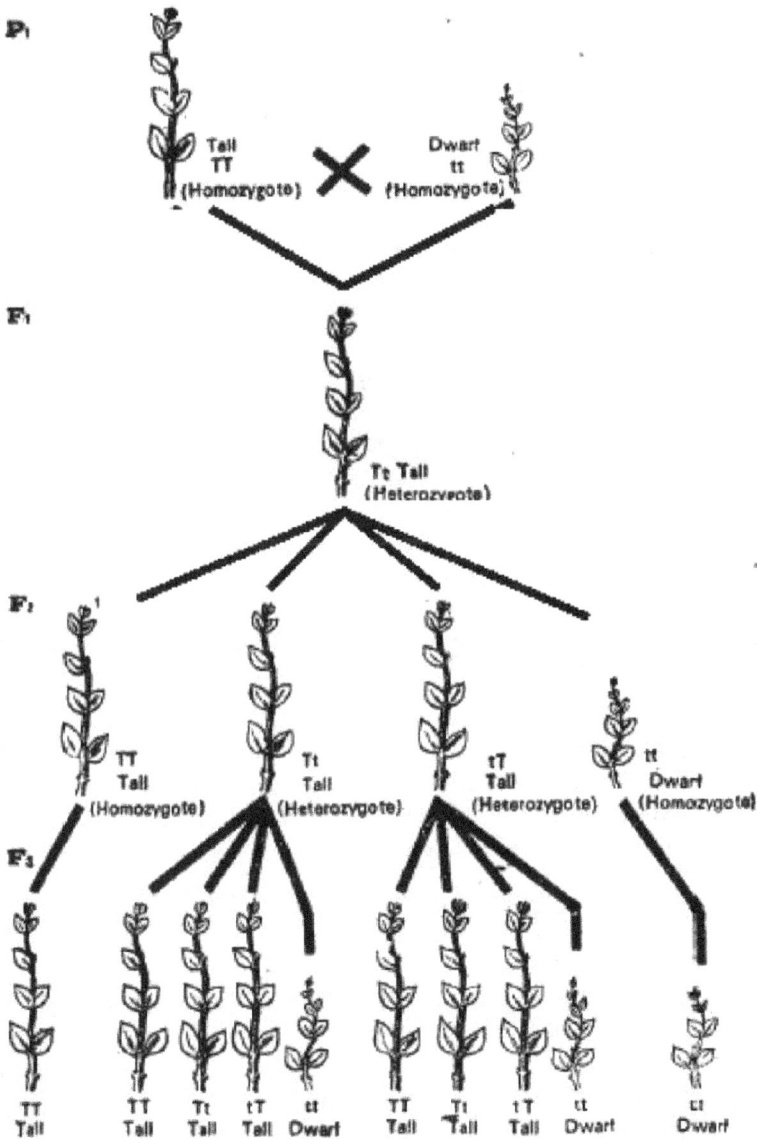

remain together throughout the life (from the zygote to the gametogenesis stage) without contaminating or mixing with each other and finally separate or segregate from each other during gametogenesis, so that, each gamete receives only one allele either dominant or recessive.

For example, the F_1 hybrids (Tt) of a monohybrid cross between tall (TT) and dwarf (tt) pea plant has one dominate allele (T) for tallness and one recessive allele (t) for dwarfness. This genotype of F_1 hybrids remains the same from the unicellular zygote stage to the gametogenesis stage of multicellular adult plant. These F_1 hybrids by selffertilization produce tall and dwarf plants in the ratio of 3 : 1. It means that tall and dwarf alleles though, remain together for long time but does not contaminate or mix with any one and both alleles segregate to produce gametes which either having dominant allele T or recessive allele t. These gametes unite to produce the 3 : 1 phenotypic ratio in F_2.

Mendel's Law of Independent Assortment

Mendel's law of independent assortment or **recombination off genes** states that when the gametes are formed the members of the different pairs of factors (genes) segregate quite independently of each other and that all possible combinations of the factors (genes) concerned will be found among the progeny. (figure on next page).

Vapour and Fluid Theory

Early Greek philosophers speculated that the hereditary informations of parents existed in the form of vapours of fluids. Pythagoras (500 B.C.) speculated that a moist "vapour" descended from the brain, nerves and other body organs of the male during the coitus and from these vapours an embryo was formed in the uterus of the female.

According to him, the male transmitted all the characters of the embryo and the female does not. However, another Greek philosopher of the same age, Empedocles thought that both parents contributed equally to the embryo and each parent produces "semen" which arises directly from various body parts.

After 200 years, another Greek philosopher Aristotle forwarded a highly imaginative speculation that the semen of the male had certain vitalizing or "**dynamic**" effect and it was supposed to be highly purified blood. According to him, the female furnished the inert building materials, while the male gives the motion and new life to the material.

Magnetic Power Theory

In the 17th century W. Harvey (1578-1657), after performing certain experiments on deer proposed the theory called magnetic power theory. He suggested that as iron by friction with a magnet possesses the magnetic properties, so that the uterus by the friction of coitus acquires some magnetic power to conceive an embryo.

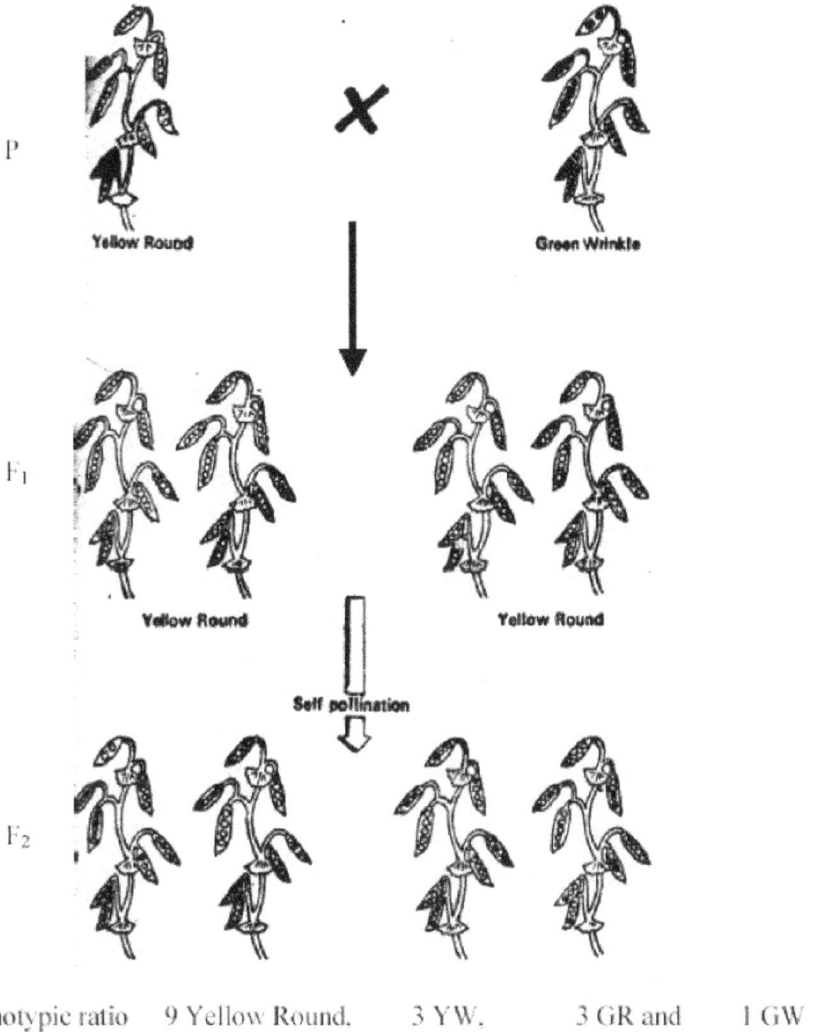

| P | Yellow Round | X | Green Wrinkle |

F₁ Yellow Round   Yellow Round

Self pollination

F₂

F₂ phenotypic ratio   9 Yellow Round,   3 YW,   3 GR and   1 GW

Preformation Theory

Leonardo da Vinci (1452-1519) revealed the fact that the male and female parents contribute equally to the heredity of the offspring. W. Harvey (1651) speculated that all living things (including man) originate form eggs and that the semen only plays vitalizing role. It was R. de Graaf (1641 – 1673) who observed that the progeny would possess the characteristics of both parents (i.e., mother and father) and therefore, suggested that both the parents should contribute to heredity (biparental inheritance). The sperms of man and other mammals were observed first of all by A.V. Leeuwenhoek in 1677. The mammalian egg was discovered by Von Baer in 1828. N. Grew (1682) first of all reported the reproductive organs of plants. With the discovery of eggs and sperms the biologists of 17th century started to speculate that the new individual was completely preformed in miniature in the gamete. According to different workers such miniature preformed embryo was speculated to occur either in the egg or sperm. Swammerdam (1637-1680), for example, thought that a tiny preformed frog occurred in the animal hemisphere of the frog egg and that became simply larger by feeding on the food stored in the vegetal hemisphere of the egg. Another biologist, Hartsoeker (1695) published a figure showing a miniature man known as mankin or homunculus in the head of the human spermatazoa. Such Preformation theories had been supported by Leeuwenhoek (1632-1723), Malphigi (1673), Reaumur, Bonnet (1720-1793), Spallazzani (1729-99) and other workers of 17th and early 18th centuries. With the development of improved microscopy and other cytological techniques in 17th and 18th centuries, it became clear to biologists that neither the egg nor the sperm contained a preformed individual but that each was a relatively uniform, homogenous mass of protoplasm.

Lamarck's Theory

The French biologist Jean Bapthiste de Lamarck (1744-1829) proposed the theory that environmental changes cause modifications in organisms and that such modifications are transmitted to subsequent generation. He believed that environment acts directly on plants and indirectly on higher animals.

Lamarck said that changes in environmental conditions create new needs in animals. Conscious efforts of the animals to adapt to the environment involve the use of certain organs, thereby causing them to become large, strong and well-developed. Other organs are not used and so become smaller, weaker and less well-developed. Such bodily changes are called acquired characters since an animal achieves them by its own exertions to adapt to the environment. Acquired characters, according to Lamarck, are then passed on to the offspring of the organism that acquired them, and new species originate by accumulation of these modification.

The giraffe dwells in the interior arid parts of Africa where there is not much herbage. According to Lamarck, the giraffe was obliged to feed on the leaves of tall trees and to strain itself continuously to reach them. Such exercise caused the necks and legs to grow in length. The increased length was inherited by the progeny, which, in turn, stretched their necks and legs and transmitted their increased length to their own offsprings. Thus, has evolved the present day six-meter high giraffe.

Detailed studies have failed to show that acquired characters are inherited. Most biologists have therefore abandoned the theory of inheritance of acquired characters, otherwise known as Lamarckism.

Darwin's Theory

In 1858, Charles Darwin (1809-1882) and Wallace independently proposed the *'Theory of Natural Selection'*. According to this theory, many more individuals of each species are born than can possibly survive and consequently there is always a struggle for existence. If hereditary differences occur within the wild species of plants, nature will eliminate some and select others.

Over-production, struggle for existence, hereditary variations and survival of the fittest are thus the important principles of the theory of natural selection.

Ten years after the publication of the *Origin of Species* (1859), Darwin adopted the doctrine of the inheritance of acquired characters but he proposed a new theory of how it happened. He modified the views of Spencer and proposed the 'Hypothesis of Pangenesis' (1868).

Darwin assumed that hereditary particles termed pangenes or gemmules, are produced by every part of the body during the life time of an organism and that, these assume the characters of the various parts of the body from which they were derived, together with whatever modifications the latter may have acquired. Eventually all the pangenes accumulate to form the germ cells which give rise to the new individual, thus ensuring the development of the parental characters and inheritance of acquired characters.

Germplasm Theory

Weismann (1834-1914), a German Zoologist, suggested in 1887 that a reduction in chromosome number took place during the formation of the egg and the sperm, and that the original number was restored when the egg and the sperm fused. In 1892, he suggested that the maternal and paternal chromosomes separated during the reduction division and that they recombined when the gametes united.

According to Weismann's *'Germplasm Theory of Heredity'*, the hereditary particles called *ids* (what we now call as genes) situated on *idants* (what we now call as chromosomes) constituted the germplasm. The germplasm is handed down form parent to offspring and it gives rise to the body or **soma** (somatoplasm) whose character it determine. The germplasm is independent of the body and whatever happens to this body has no effect on the germplasm which his contained within it.

According to Weismann, acquired characters cannot therefore be inherited. To prove this he cut off the tails of mice for twenty two generations and found that the progeny consisting of 1,592 individuals had tail of normal length. The independence of the germplasm from the somatoplasm was shown by the ovary transplantation experiment in guinea pig. Ordinarily, when albino guinea pigs are mated with albinos, only albinos are produced. Castle and Phillips removed the ovaries of an albino guinea pig and grafted in their place the ovaries of a black guinea pig. The albino animal with the ovary of the black one was then mated with an albino. All the offspring were found to be black, thereby proving that the germplasm (*i.e.,* the ovary from the black guinea pig) is not affected by the somatoplasm (*i.e.,* the body of the albino).

Mutation Theory

Charles Darwin believed that evolution is due to natural selection of small hereditary variations occurring among individuals of any species. Bateson did not agree with Charles Darwin. He believed that evolution is due to large discontinuous variations.

De Vries (1848-1935) introduced the term 'mutation' for these large, discontinuous changes in the genotypes and proposed the '*Mutation Theory*', according to which sudden hereditary changes lead to evolution.

De Vries (1901) observed that the evening primrose *Oenothera lamarckiana,* a native of America, was growing wild in Holland. In a population of this weed, he observed some plants which differed in some characters from the typical *Oenothera lamarckiana.* Since it is a self fertilized species, he felt that these variants have arisen suddenly rather than as hybrids. He transplanted them to his garden and studied them for several years. He observed that variation continued to arise spontaneously and that these variations were inherited. He called these drastic changes as mutations and maintained that mutations play an important role in the evolution of new species.

Morphology of the Eukaryotic Chromosomes

The eukaryotic chromosomes differ from the prokaryotic chromosomes in morphology, chemical composition and molecular structure. The shape of the eukaryotic chromosomes is changeable from phase to phase in the continuous process of the cell growth and cell division. They are thin, coiled, elastic, contractile, thread-like structure during the interphase (when no division of cell occurs) and are called chromatin threads.

During metaphase stage of mitosis and prophase of meiosis, these chromatin threads become highly coiled and folded to form compact and individually distinct ribbon-shaped chromosomes.

These chromosomes contain a clear zone called kinetochore or centromere along their length. The number and position of centromeres is variable, but is definite in a specific chromosome of all the cells and in all the individuals of the same specie. Thus, according to the number of the centromere the eukaryotic chromosomes may be acentric (without any centromere), monocentric (with one centromere), dicentric (with two centromeres) or polycentric (with more than two centromeres). The centromere has small granules or spherules and divides the chromosomes into two or more equal or unequal chromosomal arms. According to the position of the centromere, the eukaryotic chromosomes may be rod-shaped (telocentric and acrocentric) J-shaped (submetacentric) and V-shaped (metacentric).

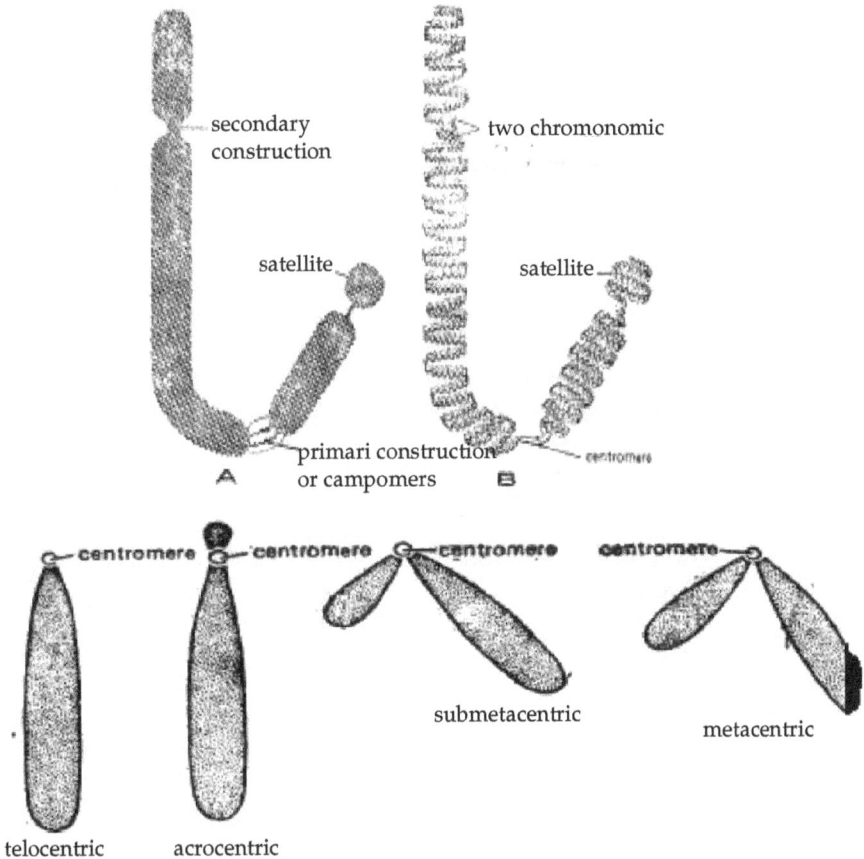

Chemical structure of chromosomes: Chemically, the eukaryotic chromosomes are composed of deoxyribonucleic acid (DNA), ribonucleic acid (RNA), histone and non-histone proteins and certain metallic ions. The most important enzymatic proteins of chromosomes are phosphoproteins, DNA polymerase, RNA-polymerase, DNA-pyrophosphorylase, and nucleoside triphosphatase. The metal ions as Ca^+ and Mg^+ are supposed to maintain the organization of chromosomes intact.

Molecular structure of chromosomes: According to the recent and widely accepted theory of Duparaw (1965, 1970) and Hans Ris (1967) called unistranded theory, each eukaryotic chromosome is composed of a single, greatly elongated and highly folded nucleoprotein fibre of 100Å thick. This nucleoprotein fibre in its turn is composed of a single, linear, double-stranded DNA molecule which remains wrapped in equal amounts of histone and non-histone proteins and variable amount of different kinds of RNA.

Kinds of chromosomes: The eukaryotic chromosomes have been classified into autosomes and sex chromosomes. The autosomes have nothing to do with the determination of sex and exceed in number than sex chromosomes. The sex chromosomes determine the sex of their bearer. They are usually two in number and are usually of two kinds: X chromosomes and Y chromosomes.

Karyotype: For cytogentical studies, when the chromosomes of a species are arranged according to their shape, size and structure, than that is called karyotype of that species.

Ideogram: When the karyotype of a species is represented by the diagram then such diagrams are called idiograms.

Special Types of Chromosomes

1. Polytene chromosomes – *e.g. Drosophila melanogaster*
2. Lampbrush chromosomes
3. B-chromosomes
4. Holokinetic chromosomes

Gene

Mendel postulated that 'Characters' are determined by 'elements' found in the sex cells or gametes. The character and its determinant are thus different and Bateson coined the word 'factor' for that which determines a character. The Danish geneticist Johannsen recognized that there is something in the gametes and in the fertilized egg that determines a character and he proposed the word '**gene**' for it.

Gene can be defined as the hypothetical unit of inheritance located at a fixed position (*i.e.,* the locus) on a chromosome which by interaction with the other genes, the cytoplasm and the environment controls the development of a character.

Allele

Allele is defined as one of a pair (or series) of forms of a gene situated at the same locus of homologous chromosomes.

Homozygote

The British geneticist Bateson introduced the term homozygote (homo = same; zygos = yolk) for an organism in which the two genes at the same locus of homologous chromosomes are identical.

Heterozygote

An individual formed by the union unlike gametes is said to be a hybrid. Bateson called this a heterozygote (hetro = different; zygos = yolk) because the two genes at the same locus of homologous chromosomes are not identical.

Genome

A haploid set of chromosome is called as genome.

Phenotype and Genotype

Johannsen clearly brought out the difference between the visible character and the invisible gene that is responsible for the character. He coined the world phenotype (phenol = appear) for the visible character of an individual and the word genotype for the heredity of a plant or animal.

Back cross

Back cross is a cross between a hybrid and either of its parents.

Test cross

Test cross is a cross between hybrid and a recessive homozygote parent.

Chromosomal theory of inheritance

Mendel's Laws of Inheritance assume that the hereditary materials are particles called genes found in the cells of all living organisms. Genes have neither been seen nor analysed chemically but it is estimated that the diameter of one gene assuming it to be a spherical particle, is something like 6 millimicrons (0.000006 millimeter). Genes are thus fundamental units of life, just like atoms which are the ultimate units of matter. Mendel established the existence of genes without knowing anything about chromosomes, in fact, several years before chromosomes had been named or described in detail. The regular and precise longitudinal division of the chromosomes into two identical halves and the distribution of the two halves to the two daughter cells by mitosis, the neat separation of chromosomes and the reduction in the number of chromosomes from the diploid (2n) to the haploid state (n) during the formation of gametes by meiosis and the restoration of the diploid number of chromosomes in the zygote by fertilization showed that the chromosomes are of great importance to the cell.

Hypothesis of Sutton and Boveri

Sutton, an American Biologist, and Boveri a German Cytologist in 1902, put the hypothesis that the Mendelian genes must be carried on the chromosomes forth simultaneously, but independently,. The chromosomes maintain their individual identity, just as do genes. In favourable materials, each pair of chromosomes can be seen to be different from every other pair. Similarly genes have individuality as can be inferred from the specific effects produced by each gene. Chromosomes are found in pairs, each member of which has been derived from one of the two parents. The facts of inheritance can be satisfactorily explained only on the assumption that genes also occur in pairs, one member of each pair being contributed by one parent and the other by the other parent.

Proofs for the theory of Sutton and Boveri

The first definite suggestion that a chromosome determines a character came from McClung, an American zoologist, when he discovered that the male grasshoppers differ from the females in the absence of one chromosome. That the X chromosome determines sex is seen from the fact that the two types of sperms differ only in that; one type has a X chromosome while the other lacks it.

Morgan's Proof for the Chromosome Theory

The discovery of sex-linked genes by Morgan in 1910 furnished another proof for the chromosomal theory of inheritance. He showed that the transmission of the recessive gene for white colour of the eye in Drosophila melanogaster depends upon the sex, which carries the gene initially. In a cross between a red-eyed female and a white-eyed male, the different results from the reciprocal crosses can be explained only on the assumption that the gene for colour of the eyes is located on the X chromosome. Morgan thus showed that the distinctive pattern of inheritance of sex-linked genes parallels the transmission of the X chromosome.

Morgan's Theory

The work of Morgan and Bridges firmly established the fact that specific genes are borne on specific chromosomes. Study of linkage and crossing over in *Drosophila melanogaster* by Morgan, Sturtevant, Muller and Bridges threw more lightly on the genes on the one hand and the chromosome on the other. From the co-ordinated genetic and cytological studies on Drosophila, Morgan postulated that genes are arranged in a linear order along the length of the chromosome, each gene having a fixed place on the chromosome, and its allele a corresponding position on the homologous chromosome. He also put forward the hypothesis that the degree of linkage depends upon the distance between the linked genes in the chromosome. This led to a new field dealing with mapping of chromosomes.

TYPES OF DOMINANCE – (*with examples*)

1. Complete dominance
2. Incomplete dominance
3. Co-dominance, and
4. Over dominance.

1. COMPLETE DOMINANCE

When a dominant allele masks completely the phenotypic expression of the recessive allele in a heterozygote, then only dominant traits takes place in the F_1 and F_2 heterozygotes.

2. INCOMPLETE DOMINANCE

When a dominant allele does not mask completely the phenotypic expression of the recessive allele in a heterozygote, then a blending of both dominant and recessive traits takes place in the F_1 and F_2 heterozygotes. This phenomenon is known as incomplete or partial dominance. **Examples:** When a homozygous red flowered pea plant is crossed with a homozygous white flowered pea plant, the F_1 heterozygotes are found to have pink flowers. When the F_1 pink flowered heterozygotes are self crossed, they produce a F_2 progeny having identical phenotypic and genotypic ratio of 1 red (RR) : 2 pink (Rr) : 1 white (rr), as has been illustrated in following figure.

The snapdragons (*Antirrhinum majus*) and four-o'clock plants (*Mirabilis jalapa*) also provide good examples of incomplete dominance.

3. CO-DOMINANCE

In the phenomenon of co-dominance, both dominant and recessive alleles lack their dominant and recessive relationships and both have capability to express them phenotypically, in the heterozygous condition. In a heterozygote of co-dominant nature, the dominant and recessive traits occur side by side. The F_1 heterozygotes produce a F_2 progeny in the phenotypic and genotypic ratios of 1 : 2 : 1 like the incomplete dominance. **Example:** The best example of co-dominance is found in cattles. When a white-coated cattle is crossed with a red-coated cattle, the F_1 heterozygote are found to have a phentoype of reddish gray or roan colour. The roan coat of a F_1 heterozygote has no hair of intermediate colour between red and white, but rather has a mixture of red hair and white hairs. The F_1 heterozygotes produce a F_2 progeny of phenotypic and genotypic ratios of 1 : 2 : 1.

4. OVER DOMINANCE (HETERO DOMINANCE)

When the heterozygotes have a more extreme phenotype than either of the corresponding homozygotes (homozygous parents), then it is usually referred to as overdominance, super-dominance or hetero-dominance (Serra, 1959). **Example:** There is heterodominance, when the heterozygote Aa between a pair of factors; which control size is bigger than the homozygotes AA or aa. This type of allelic relation which implies interaction between the alleles, or of these with other factors of the genotype, may be found in quantitative characters and especially those such as size, production, vigour, etc., which are of importance in the breeding of animals and plants.

Lethal Gene

The term lethal is applied to those changes in the genome of an organism, which produce effects severe enough to cause death. Lethality is a condition in which death of a certain genotype occurs prematurely. The fully dominant lethal allele kills the carrier individual both in its homozygous and heterozygous conditions. It occasionally arises by mutation from a normal allele. The individuals with a dominant lethal allele die before they can produce the progeny. Therefore, the mutant dominant lethal allele is removed from the population in the same generation in which it arose. The recessive lethal allele kills the carrier individual only in homozygous condition. They may be of two kinds: (i) one which has no obvious phentoypic effect in heterozygotes and (ii) one which exhibits a distinctive phenotype when in heterozygous condition. The lethal alleles modify the 3 : 1 phenotypic ratio into 2 : 1. **Example:** In mice.

Penetrance and Expressivity

PENETRANCE: The ability of a given gene or gene combination to be expressed phenotypically to any degree is called penetrance.

In other words, penetrance refers to the statistical regularity with which a gene produces its effect when present in the requisite homozygous (or heterozygous) state. It is of following two kinds:

1. **Complete Penetrance :** In pea, the alleles (RR) for red flowers and the alleles (rr) for white flowers have complete penetrance in homozygous conditions.

2. **Incomplete Penetrance :** Polydactyly in man.

   Interaction of GENES:

   When different pairs of alleles influence the same character of an individual, it is likely that the expressions of these genes interact. As two different genes interact and affect the same character, such a genetic interaction is said to be intergenic or nonallelic. In nonallelic interactions different genes located on the same or different chromosomes interact with one another for the expression of a single phenotypic trait of an organism.

Intergenic or non-allelic interactions may suppress or mask the action of a gene at another locus or modify partially or completely the effect of another gene. This nonallelic interaction is otherwise called *epistasis.*

Def'n: A kind of interaction between genes belonging to different pairs of alleles, the dominant allele in one of the pairs preventing the dominant allele in the other pair from expressing itself. Thus, the gene A may be epistatic over B. B is then said to be hypostatic to A.

Intergenic Non-epistatic interaction (9: 3 : 3 : 1 Ratio)

Each breed of poultry possesses a characteristic type of comb.

Types of Epistasis

1. Dominant epistasis (12:3:1)

2. Recessive epistasis (9:3:4)

3. Duplicate dominant epistasis (15:1)

4. Duplicate recessive epistasis (9:7)

5. Dominant and recessive epistasis (13:3)

6. Duplicate genes with cumulative effect (9 : 6 : 1)

Summary of epistatic ratios

The epistatic ratios can be summarised as under:

	1 2 3 4 5 6 7 8 9 10 11 12 13 14 15 16			
Genotypes	A - B -		A - b b	a a B -	aabb
Independent assorment/ Non-epistasis	9		3	3	1
Dominant epistasis	12			3	1
Recessive epistasis	9		3	4	
Dominant and Recessive interaction	13			3	
Duplicate dominant interaction	15				1
Duplicate recessive interaction	9			7	
Duplicate genes with cumulative effect	9			6	1

Summary of epistatic ratios.

Multiple Alleles

When any of the three or more allelic forms of a gene occupy the same locus in a given pair of homologous chromosomes they are said to constitute a series of multiple alleles. In other words, all the mutant forms of a single wild type gene constitute a series of multiple alleles.

Characters of Multiple Alleles

1. Multiple alleles of a series always occupy the same locus in the chromosome.

2. Because, all the alleles of multiple series occupy the same locus in chromosome, therefore, no crossing-over occurs within the alleles of a same multiple allele series.

3. Multiple alleles always influence the same character.

4. The wild type allele is nearly always dominant, while the other mutant alleles in the series may show dominance or there may be an intermediate phenotypic effect.

5. When any two of the mutant multiple alleles are crossed, the phenotype is mutant type and not the wild type.

    Multiple Allelic inheritance of A, B, AB and O Blood Types: Bernstein (1925) proposed that inheritance of A, B, AB and O blood types of man is determined by a series of three allelomorphic genes. The L gene exists in three different allelic forms: LA, LB and LO.

    Nilsson-Ehle's studies on kernel colour in wheat

    The Swedish geneticist Nilsson-Ehle (1908) effect crosses between different true breeding strains of wheat with red kernels and those with white kernels. Careful examination revealed that all the red kernels were not of the same intensity of colour. It was possible to separate the F_2 into the following classes:

    Dark red 1

    Medium dark red 4

    Medium red 6

    Light red 4

    White 1

    It is evident that red colour is due to two pairs of alleles. Each gene is capable of producing red colour. Each is incompletely dominant over white and is cumulative in its effect. The intensity of the red colour depends upon the number of colour producing genes present. Dark red is due to the presence of four contributing genes for red, medium dark red to three contributing genes, medium red to two contributing genes and light red to one contributing gene.

The F_2 ratio in wheat

Genotype	Genotypic ratio	Phenotype
R1R1R2R2	1	Dark red
R1R1R2r2	2	Medium dark red
R1r1R2R2	2	Medium dark red
R1r1r2R2	4	Medium red
R1R1r2r2	1	Medium red
r1r1R2R2	1	Medium red
R1r1r2r2	2	Light red
r1r1R2R2	2	Light red
r1r1r2r2	1	White

Differences between polygenic and oligogenic traits

S.N.	Polygenic traits	Oligogenic traits
1.	Governed by several genes	Governed by few genes
2.	Effects of each gene is not detectable	Effect of each gene is detectable
3.	Usually governed by additive genes	Governed by non-additive genes
4.	Variation is continuous	Variation is discontinuous
5.	Separation into different classes is not possible	Separation into different classes is possible
6.	Highly influenced by environmental	Little influence by environmental factors
7.	Statistical analysis is based on mean, variances and covariance	Statistical analysis is based on frequencies or ratios.

Linkage group

The phenomenon of inheritance of linked gene in same linkage group is called linkage.

The number of linkage groups will be equal to the haploid number of chromosomes which the species possesses. Thus, maize which has 10 pairs of chromosomes has 10 linkage groups.

- **Views of classical geneticists on linkage**

Coupling and Repulsion Hypothesis of Bateson and Punnett

Bateson and **Punnett** (1905-1908) formulated the 'hypothesis of coupling and repulsion' to explain the unexpected F_2 results of a dihybrid cross between a homozygous sweet pea (*Lathyrus odoratus*) having a dominant alleles for blue or purple flowers (RR) and long pollen grains (Ro Ro) with another homozygous double recessive plant (rr, roro) with red flowers and round pollen grains. When they test crossed a heterozygous blue or purple long (Rr, Roro) plant with recessive parent (rr, roro), besides getting the 1 : 1 : 1 : 1 test cross ratio, they received phenotypic ratio of 7 : 1 : 1 : 7.

- **Morgan's views on Linkage** : Bateson and Punnett could not explain the exact reasons of coupling and repulsion, and it was T.H. Morgan who while performing experiments with *Drosophila*, in 1910, found that coupling or repulsion was not complete. He further suggested that the two genes are found in coupling phase or in repulsion phase, because they are present on the same chromosome (coupling) or on two different homologues chromosomes (repulsion). Such genes are then called linked genes and the phenomenon of inheritance of linked genes is called linkage by Morgan.

- **Chromosome Theory of Linkage :** The chromosome theory of linkage of Morgan and Castle states that:

  (i)   The genes which show linkage are situated in the same pair of chromosomes.

  (ii)   The linked genes remain arranged in a linear fashion on the chromosome. Each linked gene has a definite and constant order in its arrangement.

  (iii) The distance between the linked genes determines the degree of strength of linkage. The closely located genes show strong linkage then the widely located genes which show weak linkage.

  (iv)   The linked genes remain in their original combination during the course of inheritance.

Kinds of Linkage: The phenomenon of linkage is of following two kinds:

  1. **Complete Linkage:** Example: According to Bridge all the genes of male *Drosophila* remain completely linked. Further, in a mutant strain of *Drosophila,* the genes for bent wings (b+) and shaven bristles (svn) of the fourth chromosome exhibit complete linkage.

  2. **Incomplete Linkage:** Example: Incomplete linkage has been observed in pea, *Zea mays* (maize), tomato, female (*Drosophila,* Mice, poultry, and man. Here, the examples of linkage have been considered only for Drosophila and *Zea mays* (maize).

Linkage groups

All the linked genes of a chromosome form a linkage group. Because, all the genes of a chromosome have their identical genes (allelomorphs) on the homologous chromosome, is considered as one. The number of linkage groups of a species thus corresponds with haploid chromosome number of that species.

Examples:

1. Drosophila has 4 pairs of chromosomes and 4 linkage groups.

2. Man has 23 pairs of chromosomes and 23 linkage groups.

*   **Significance of Linkage**

    The phenomenon of linkage has one of the great significance for the living organism that it reduces the possibility of variability in gametes unless crossing over occurs.

*   **Crossing over**

    The process of crossing over can be defined as "a process which produces new combinations (recombinations) of genes by interchanging of corresponding segments between non-sister chromatids of homologous chromosomes."

    According to its occurrence in the germinal or somatic cells following two types of crossing over have been recognised :

    A. Germinal or **meiotic crossing over:** Commonly crossing over occurs only in the germinal cells of reproductive organs during the process of gametogenesis which includes meiosis. This type of crossing over is called germinal or meiotic crossing over. It is universal in its occurrence and has great genetic significance.

    B. Somatic or mitotic crossing over: Sometimes crossing over may occur during mitosis of somatic cells. This type of crossing over occurs in rare cases, has no genetic significance and is called somatic or mitotic crossing over.

*   **Mechanism of meiotic crossing over:** Process of crossing over include following stages, *viz., synapsis, duplication of chromosome, crossing over and terminalization* (Whitehouse and Hastings, 1965).

1. Synapsis

During zygotene stage, of prophase-I of meiosis occurring in developing sex cells, the homologous chromosomes come close to each other and pairing on Synapsis between the homologous chromosomes (genetically identical chromosomes) takes place. It is started during zygotene when homologous chromosomes are held to make contact with each other at one or more points from which synapsis extends into adjacent regions and it ends or reaches its maximum in pachytene after which the homologues fall apart except the regions of chiasmata. The resultant pairs of homologous chromosomes are called bivalents. Synaptinemal Complex-Montrose J. Moses (1955) has revealed a highly organized structure of filaments called synaptinemal complex in between the paired chromosomes of zygotene and pachytene stages in crayfish by electron microscopy.

2. Duplication of Chromosomes

The synapsis is followed by duplication of chromosomes. During this stages, each homologous chromosome of a bivalent splits longitudinally and form two identical sister chromarids, so that, each bivalent is now, composed of four chromatids. A bivalent having four chromatids is called tetrad.

3. Crossing over

The crossing over occurs in the homologous chromosomes only during the four stranded or tetrad stage. During the process of crossing over, two non-sister chromatids first break at the corresponding points due to the activity of a nuclear enzyme called endonuclease (Stern and Hotta, 1969). Then a segment on one side of each break connects with a segment on the opposite side of the break, so that the two non-sister chromatids cross each other at the point of break and exchange. The fusion of chromosomal segments with that of opposite one takes place due to the action of an enzyme called ligase (Stern and Hotta, 1969). According to the recent findings a little amount of DNA synthesis takes place during the crossing over process and that little amount (about 3% of total genome) of DNA is thought to repair the broken chromosomes. The crossing of two chromatids is called chiasma formation and the resultant cross as chiasma or chiasmata. The crossing over thus, includes the breaking of chromatid segments, their transposition and fusion.

4. Terminalisation

After the completion of crossing over, the non-sister chromatids start to repel each other because the force of synapsis attraction between them decreases. The chromatids separate progressively from the centromere towards the chiasma and the chiasma itself moves in a zipper fashion towards the end of tetrad. The movement of chiasma is called terminalisation. Due to the terminalisation the homologous chromosomes are separated.

- **Cytological Proof for crossing over**

The first cytological demonstration of genetic crossing over has been given by Stern (working with Drosophila) and H.B. Creighton and B. McClintock (working with maize) in 1931.

- **Significance of Crossing over**

1. The frequency of crossing over is of great use in constructing genetic maps of the chromosomes.

2. It provides direct evidence for linear arrangement of linked genes in chromosomes.

3. It increases the frequency of genetical variation which are the raw materials of organic evolution.

- **Strength of linkage and recombination**

  Two point: Crossing over between two linked genes is calculated by test crosses.

  Three point test cross: As three point test cross or trihybrid test cross (involving three genes).

- **Interference and Coincidence**

  Interference : "The tendency of one crossover to interfere with the other crossover is called interference."

The net result of this interference is the observation of fewer double crossover types than would be expected according to map distances the strength of interference varies in different segments of the chromosome and is usually expressed in terms of a coefficient of coincidence, or the ratio between the observed and the expected double crossovers.

$$\text{Coefficient of coincidence} = \frac{\% \text{ of observed double crossovers}}{\% \text{ of expected double crossovers}}$$

The coincidence is the complement of interference, so: When interference is complete **(1.0)**, no double crossovers will be observed and coincidence becomes **zero**. When, interference decreases, coincidence increases. Coincidence values ordinarily vary between **0 and 1**. Coincidence is generally quire small for short map distance. There is no interference across centromere.

Sex determination and Sex linkage

Sex differentiation in living organisms into male and female cause's morphological, physiological and behavioural differentiation between the two sexes and this phenomenon is called sexual dimorphism.

The precise form of the chromosomal differences between the sexes is not the same in different organisms. Four types of sex chromosome mechanism or are as following:

Sex chromosome mechanism

(a) **Heterogametic Male**

XX-XO type: In squash bug, *Protenor,* the females have 14 chromosomes and the males have only 13 chromosomes in their somatic cells.

XX-XY type: In *Drosophila melanogaster* the female has four pairs of chromosomes as follows: (1) a pair of rod-shaped chromosomes, (2) a pair of V-shaped chromosomes, (3) a pair of slightly longer V-shaped chromosomes, and (4) a pair of very short dot-like chromosomes. This type of sex determination in which the female has two X chromosomes and the male one X and the Y chromosome is very widespread, being found in many invertebrates including insects, in some fishes, in mammals including man and in many dioecious plants like *Melandrium album, M. rubrum, Humulus lupulus, Rumex angiocarpa, Salix, Smilax, Cannabis* and *Populus* and also in human beings.

(b) **Heterogametic Female**

ZO-ZZ type: In all the above instances, the female is the homogametic sex; because it produces eggs, all of which are alike and the male is the heterogametic sex because it produces two kinds of sperms. But there are instances where the female is the heterogametic sex and the male is the homogametic one.

In a moth, *Talaeoporia,* the females have 59 chromosomes and the males have 60 chromosomes in their somatic cells. The eggs are of two kinds, one kind with 29 chromosomes and the other kind with 30 chromosomes.

ZW-ZZ type: In birds, including the domestic fowl, certain insects, fishes and reptiles, the female has an unlike pair of chromosomes, ZW, and forms eggs of two sorts, one with a 'W' chromosome and the other with a 'Z' chromosome. Among plants, *Fragaria elatior* is one in which the female is ZW and the male is ZZ.

Balance theory of sex determination: (comes from the work of Bridges (1921) on Drosophila.)

To quote Bridges, 'Both sexes are due to the simultaneous action of two opposed sets of genes, one set tending to produce the characters called female and the other to produce the characters called male'. Which sex actually develops is decided by the balance, *i.e.,* by the preponderance of the female determining or of the male-determining genes.

Sex-influenced character

All genes which are carried by the chromosomes are said to be sex-linked. All known sex-linked genes lead to phenotypes which have nothing to do with sex. Sex-influenced characters, are characters which may be expressed differently in the two sexes even when their genotypes are identical. The most common expression of sex-influenced is that dominance is reversed between the sexes. Genes determining sex-influenced characters are borne on autosomes, *e.g.,*

(i) A typical example of a sex-influenced character is the presence of horns in sheep.

(ii) Baldness in human beings is a sex-influenced character which is recessive in females and dominant in males.

Sex-limited character

Sex-limited inheritance is an extreme type of sex-influenced in which a particular phenotype can be expressed only in one sex. Unlike sex-influenced characters in which gene is dominant in one sex and recessive in the other, sex-limited characters are controlled by genes which have no visible influence at all in one sex either as a homozygote or as a heterozygote, *e.g.,* in domestic poultry, **cock-feathering** is a character limited to the male sex.

Sex reversal

In several species of plants that are normally bisexual, suppression of the male or female structures has been observed in nature. The androecium getting converted into petels in ornamental plants or carpels as in carrot and cabbage or pistils as in maize, papaya and primrose has been observed. The phenomenon in which there is suppression of one sex at the expense of the other is called sex reversal. The sex reversals are mostly due to physiologicl and biochemical alterations involving sex hormones, *e.g.,* in maize, rarely it is observed that the male inflorescence called tassel bears seeds due to sex reversal. The recessive gene 'ba' is responsible for barren plants and another recessive gene 'ts' is responsible for tassel seed. Sex reversal in maize is due to the genetic constitution of the plants.

Cytoplasmic inheritance

Cytoplasmic inheritance is due to the plasmagenes located in cell organelles (plastids and mitochondria). The characteristic features of this inheritance are summarized below.

- **Characteristics of Cytoplasmic Inheritance**
    1. Reciprocal Differences
    2. Lack of segregation
    3. Irregular Segregation in Biparental Inheritance
    4. Somatic Segregation.
    5. Association with Organelle DNA
    6. Nuclear Transplantation.
    7. Transfer of Nuclear Genome Through Backcrosses.
    8. Mutagenesis.
    9. Lack of Chromosomal Location
    10. Lack of Association with a parasite, Symbiont or Virus.

These cytoplasmic extra-nuclear genes or DNA molecules of plasmids, mitochondria, chloroplasts, endosymbionts and cellular surfaces have a characteristic pattern of inheritance which does not resemble with that of genes of nuclear chromosomes soma, uniparental, maternal, extra-chromosomal, cytoplasmic and extranuclear inheritance.

Extra-nuclear inheritance by cellular organelles: Chloroplasts and mitochondria are organells that contain their own DNA and protein-synthesizing apparatus.

(a)   Chloroplasts inheritance in variegated four O' clock plant.

(b)   Maternal inheritance by 'iojap' plastid gene of corn.

(c)   Cytoplasmic male sterility in maize, sorghum, cotton etc.

(d)   Mitochondrial inheritance in Petite mutants of Yeast.

DNA, the genetic material

Considering the proportion of different constituents of cell, nucleic acid was found to be constant in volume in all the cells as compared to other cellular contents and hence it was inferred to be the hereditary material. There are two types of nucleic acid, the deoxyribonucleic acid (DNA) and the ribonucleic acid (RNA). By staining nucleic acid, Feulgen (1924) found that the DNA was localized in the nucleus, while the RNA was found to occur outside the nucleus in the cytoplasm.

- The experiments of Griffith (1928) with the pneumonia bacterium revealed 'transforming principle' that transformed the hereditary property of avirulent R II to virulent S III. This phenomenon is called 'Griffith effect' or 'Bacterial transformation'.

- **DNA as the genetic material in viruses:** Hershey and Chase (1952) provided direct proof that DNA is the genetic material in certain bacterial viruses.

  [Bacteriophage is a virus that infects or feeds on certain specific bacteria. **T2** bacteriophage that infects the colon bacteria, *Escherichia coli* was involved in the studies. The viral DNA was labeled with P32 and the viral capsid (protein coat) with S35, since DNA contained P and the capsid protein contained S. then the labeled viruses were allowed to infect unlabelled *E. coli* and get multiplied.]

- The interpretation of results by Avery, MacLeod and McCarty (1944) confirmed the DNA as the hereditary material.

Structure of DNA – Watson and Crick Model

Chemical composition of DNA

DNA is a complex macromolecular or polymeric chemical compound which contains four kinds of monomers (small building blocks) called Deoxyribonucleotides. Each deoxyribonucleotide is made up of (1) a phosphoric acid molecule, biologically called phosphate, discovered by Levene (1910), (2) a pentose sugar called 2-deoxyribose, also discovered by Levene (1910), and (3) four major kinds of nitrogen bases, two heterocyclic and two ringed purines, adenine (A) and guanine (G) and two one ringed pyrimidines, cytosine (C) and thymine (T), discovered by Fischer (1880).

Double Helical Model of DNA

Based on the findings of Chargaff (1950) that the total amount of purines equalled the total amount of pyrimidines (A + G = T + C), that the amount of adenine equalled the amount of thymine (A = T) and the amount of guanine equaled the amount of cytosine (G = C) and, that the ratio between total purines and total pyrimidines was always not far from one,

$$(A + G) : (T + C) = 1,$$

as well as the crystallographic evidences and X-ray differentiation photographs (Astbury, 1947, Wilkins and Franklin, 1953), the double helical model of DNA was constructed by Watson, an American biologist and Crick, a British Physicist in 1953.

The DNA molecule was conceived as a two stranded structure coiled like a rope, and hence called plectonemic, so that if the ends are permitted to revolve freely, the complementary strands could easily separate. The coil was proposed to be helical and conceived to resemble a circular staircase, maintaining the same diameter through out and having a constant width between steps. The steps are connected on either side by a railing.

The helix has a diameter of 20Å and makes a complete turn at every 34Å along its length. The distance between nucleotides is 3.4Å. Each complete turn has a stack of 10 nucleotides. The helix contains two polynucleotide chains or two stacks of 10 nucleotides each per turn.

Each complementary strand is only half the circular staircase, either side consisting of approximately half the width of the step. Each half step is connected by a railing or backbone. The railing consists of phosphate – sugar linkages which are repeated without change. The half step of one strand extends to meet the half step of the complementary strand. Each half step has either a purine or pyrimidine base. Each step consisting of two half steps is together called base pair.

The fit between the bases is determined by hydrogen bonding. The bonding involves the ability of the H atom with positive charge (H+) to be placed between an O atom with weak negative charge (O-) and a N atom with a light negative charge (N-) from opposite strands. Adenine pairs with thymine with two H bonds (A = T) and guanine with cytosine with three H bonds (G = C). These N bases are connected to each other by deoxyribose and phosphoric acid. Hydrogen bonds are generally weaker than other chemical bonds. But there are several of them, two between A and T (A = T) and three between G and C (G = C) that give rigidity and stability to the molecules.

Protein Synthesis

Central Dogma of Molecular Biology

The process of protein synthesis involves one of the central dogma of molecular biology, postulated by Crick (1958) according to which genetic information flows nucleic acid to protein.

Protein synthesis involves two steps, *viz.,* transcription and translation.

Transcription involves a sequential flow of information from DNA to mRNA. This does not involve a change of code since DNA and mRNA are complementary.

Translation involves a change of code from nucleotide sequences to amino acid sequences.

Generally the flow of information is one way, from DNA to RNA and from RNA to protein.

$$DNA \xrightarrow{\text{Transcription}} RNA \xrightarrow{\text{Translation}} Protein$$

In certain viruses, the existence of an enzyme 'RNA dependent DNA polymerase' (also called inverse transcriptase) was reported and this enzyme could synthesize DNA from a single stranded RNA template. This finding of Baltimore (1970) and others give rise to the concept of 'central dogma reverse'. According to this, the sequence of information flow is not necessarily from DNA to RNA to protein, but can also take place from RNA to DNA.

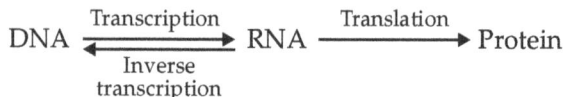

$$DNA \underset{\substack{\text{Inverse} \\ \text{transcription}}}{\overset{\text{Transcription}}{\rightleftarrows}} RNA \xrightarrow{\text{Translation}} Protein$$

Transcription

The process by which the information in the nucleotide sequence of DNA is transferred to a complementary sequence of RNA is known as transcription. The locations of transcription are (1) the nucleolus where genes rRNA are transcribed and (2) the remaining chromatin where mRNA (mRNA) is transcribed.

Translation

As soon as the mRNA is formed, it leaves the nucleus and reaches the cytoplasm where translation takes place.

Before the process of protein synthesis, the ribosomes occur in dissociated and inactive state. The mRNA binds with 30 S ribosomal subunit in the presence of a protein factor called Initiation Factor (IF). The mRNA carries triplet codons for the synthesis of proteins. Protein synthesis involves mRNA, ribosomes, amino acids and their specific tRNAs.

Translation process involves the following steps

1. Attachment of mRNA with 30 S ribosomes and formation of polyribosomes,
2. Activation of the amino acids
3. Attachment of activated aamino acid to tRNA
4. Initiation of the polypeptide chain
5. Termination of the polypeptide chain

Thus a polypeptide chain with a specific series of amino acids is formed which results in synthesis of a specific protein that involves in a specific phenotypic expression in the organism.

Genetic Code

In the DNA and RNA, there are four types of nucleotides or bases *viz.*, A, G, T, C and A, G, U, C respectively. If three bases together code for an amino acid, then $4^3 = 64$ amino acids could be coded. As the essential amino acids in a biological system are 20 in number, the possibility of one or two bases coding for each amino acid is remote.

{Nirenberg (1961), Khorana (1964) and others lead to the construction of a complete genetic code dictionary.}

Regulation of gene expression

Gene expression refers to manifestation of a phenotypic character by the activity of gene. A gene expresses itself by producing proteins or enzymes. During gene expression, there is flow of genetic information from DNA to protein. Gene expression involves two steps – transcription and translation. The genes, which are not functioning, are said to be switched off and the genes, which are functioning, are said to be switched on.

Regulation of Gene Expression in Prokaryotes

It is well studied in the bacterium, *Escherichia coli* and other prokaryotes the gene expression is regulated at two levels. They are (i) Regulation of enzyme (ii) Regulation of Transcription.

I. Regulation of Enzyme

The regulation of enzymes is brought about by the following mechanisms.

1. Enzyme induction

2. Enzyme repression

3. Feed back inhibition

II. Regulation of Transcription

Transcription refers to the synthesis of mRNA from DNA. It is a stage in gene expression. In prokaryotes, the genes are regulated at the level of transcription.

1. Negative regulation,

2. Protein regulation,

3. Auto regulation,

4. Coordinate regulation.

Operon Hypothesis

The operon hypothesis was put forward by Jacob and Monod in 1961 Nobel Prize in 1965.

Operon is a set of closely linked genes regulating a metabolic pathway in prokaryotes.

Lac Operon

1. Lac operon is a set of genes responsible for the metabolism of lactose in *E. coli*.

   (Jacob and Monod in 1961.)

2. The Lac operon consists of 3 structural genes namely Z, y and a and 3 control genes promoter gene (P), a regulator gene (I) and operator gene (O).

3. The structural genes are responsible for the synthesis of three enzymes namely, â – galactosidase by the gene Z, galactoside permease by the gene y, thiogalactoside transacetylase by the gene a.

4. The operator gene is closely linked to the first structural gene Z. when the operator gene is active, the structural genes synthesis enzymes.

5. A repressor protein synthesized by a regulator protein decides the activity of the operator gene.

6. When the repressor binds to the operator gene, the operator gene is made nonfunctional. This state of the operator gene is called repressed state and the phenomenon is called repression. During repression, the enzymes are not synthesized.

7. When lactose is introduced into the medium, it diffuse into the cell and binds to the repressor protein to form an in active inducer repressor complex.

8. In the lac operon system, lactose functions as an inducer for the synthesis of three enzymes. Hence the lac operon system is called in inducible system. The lac operon is a system of negative regulation. In negative regulation the regulator protein repressor prevents gene transcription.

Modern Concept of Gene

Fine Structure of Gene: The hereditary units, which are transmitted from one generation to the other generation, are called as genes.

Classical Definition of Gene

A gene is a unit of physiological function that occupies a definite locus in the chromosome and is responsible for a specific phenotypic character, e.g., vestigial or long wings and white and yellow eyes in *Drosophila*.

A gene is a unit of transmission or segregation because it can be segregated and exchanged at meiosis by way of crossing over.

A gene is a unit of mutation because by a spontaneous or induced change it can give rise to different phenotypic expression.

Mordern Definitions of Gene

1. **Cistron :** The portion of DNA specifying a single polypeptide chain, synonym for the tern gene of physiological function.

   Cistron was coined by Seymour Benzer. Haemoglobin therefore would require two cistrons fraction globin protein fraction one each for á and â chains.

2. **Muton :** Benzer coined the word muton to that smallest length of DNA capable of mutational change. Different formes of a mutationality-defined gene are called as homoalleles.

3. **Recon :** A recon is the smallest unit of DNA capable of recombination or of being integrated by transformation in bacteria. Recombinationally separable forms of a cistron are called heteroalleles.

"The gene of function is the sequence of nucleotides which specifies the amino acid sequence of a specific polypeptide chain."

Split Genes

There are some genes which are different from normal genes either in terms of their nucleotided sequences or fractions. Split genes have two types of sequences viz., normal sequences and interrupted sequences.

1. **Normal sequence:** This represents the sequence of nucleotides, which are included in the mRNA which is translated from DNA of split gene. These sequences code for a particular polypeptide chain and are known as exons.

2. **Interrupted sequences:** These are called as introns. These donot codes for any peptide chain. Interrupted sequences are not included into mRNA which is transcribed from DNA of split genes.

Complementation Test

The introduction of two independently occurred mutations into the same cell for the purpose of determining whether the mutation occurred in the same gene. This test can be accomplished with by mating two homozygous organisms or by somatic cell fusion.

SPECIAL TYPES OF CHROMOSOMES

1. **Polytene chromosomes :** The nuclei of the salivary gland cells of the larvae of *Drosophila melanogaster* have unusually long and wide chromosomes, 100 or 200 times in size of the chromosomes. Since the salivary gland cells do not divide after the glands are formed, yet their chromosomes replicate several times (a process called endomitosis) and become exceptionally giant – sized to be called polytene chromosomes. The polytene chromosomes of the salivary gland cells of *D.* contain 1000 to 2000 chromosomes. If the polytene chromosomes of dipteran larval salivary glands are examined at several stages of development; it is seen that specific areas (sets of bands) enlarge or "puffs".

2. **Lampbrush Chromosomes :** In diplotene stage of meiosis, the yolk-rich oocyte of vertebrates contain the nuclei with many lampbrush-shaped chromosomes of exceptionally large size. The lampbrush chromosomes (discovered by Rucker in 1892) are formed during the active synthesis of mRNA molecules for the future use by the egg during cleavage when no synthesis of mRNA molecules is possible due to active involvement of chromosomes in the mitotic cell division.

3. **B-Chromosomes:** Many plant (maize) and animal (insects and small mammals) species, besides having autosomes (A-chromosomes) and sex-chromosomes posses a special category of chromosomes called B-chromosomes without obvious genetic function. These B-chromosomes usually have a normal structure, are somewhat smaller than the autosomes.

4. **ISO Chromosomes:** An isochromosome is a chromosome in which both arms are identical. It is thought to arise when a centromere divides in the wrong plane, yielding two daughter chromosomes, each of which carries the informations of one arm only but present twice. The isochromosomes are formed during mitosis and meiosis.

VARIATION IN CHROMOSOME STRUCTURE

Types of Chromosomal Aberrations

A. Intrachromosomal aberrations

B. Interchromosomal aberrations.

A. Intrachromosomal Aberrations

When aberrations remain confined to a single chromosome of a homologous pair, they are called intrachromosomal or homosomal aberrations.

1. **Deficiencies (Deletions):** In deletion or deficiency type intrachormosomal aberration a chromosomal lacks either in an interstitial or terminal chromosomal segment, which may include only a single gene or part of a gene.

   Genetic Significance of Deficiencies— Lethal effect.

2. **Duplications (Additions):** Duplication occurs when a segment of the chromosome is represented two or more times in a chromosome of a homologous pair.

GENETIC SIGNIFICANCE OF DUPLICATIONS

1. The duplications of chromosomes are not deleterious to the organism like the deficiency, but, they usually protect the organism from the effect of a deleterious recessive gene or from an otherwise lethal deletion.

2. Large duplications can reduce the fertility as a result of meiotic complication, and in this way reduce their own probability of survival (Sybenga, 1972).

3. Relocation of chromosomal material without altering its quantity may result in an altered phenotype, this is called position effect.

INVERSIONS

An inversion is an intra-chromosomal aberration in which a segment is inverted 180 degrees.

The inversions are of following types:

(i) Pericentric inversions – When the inverted segment of chromosome includes or contains centromere, then such inversions are called heterobranchial or pericentric inversions.

If crossing over occurs with in the loop of a pericentric inversion, the resulted chromatids include half on the chromatids with duplications and deficiencies forming nonfunction. The other half of the chromatids forms functional gametes: ¼ gametes have normal chromosome order; ¼ gametes have the inverted arrangement.

(ii) Paracentric inversions – When the inverted segment includes no centromere and the centromere remains located outside the segment, then such type of inversion is called homobranchial or paracentric inversion.

Crossing over within the inverted segment of a paracentric inversion, produces a dicentric chromosome contains two centromeres and forms a bridge from one pole to the other during first meiotic anaphase. When anaphase chromosomes separate towards poles, this bridge breaks somewhere along its length and the resulting fragments contain duplications and/or deficiencies. The acentric chromosome because lacks in centromere and fails to move to both pole and so, is not included in the meiotic products. Such, breakage-fusion bridge cycles of crossing over of paracentric inversions are most common in maize. The meiotic products include half non-functional, ¼ functional normal and ¼ functional inverted chromosomes.

Genetic Significance of Inversions

(i)   Simple inversions do not have primary phenotypic effects other then on chromosome shape. Frequently, however, some DNA at a break point has been damaged and this may result in an observable mutation, often recessive (*e.g.,* c 1B lethal mutation in Drosophila).

(ii) Due to inversion a peculiar kind of position effect occurs. The transfer of a gene causes the position effect from a euchromatic segment to the vicinity of heterochromatic segment. Heterochromatinization may then extend into a displaced, originally euchromatic region and suppress the transcription of the gene in it.

(iii) Normal linear pairing is not possible in inversion heterozygotes. The difficulties encountered with pairing cause a reduction of exchange (crossing over) in and around the inversion.

(iv) They maintain heterozygosity from generations to generations.

B. INTERCHROMOSOMAL ABERRATIONS

When breaks occur in non-homologous chromosomes and resulting fragments are interchanged by both of the non-homologous chromosomes, the inter-chromosomal or heterosomal aberrations occur. The inter-chromosomal aberration is of following type:

Translocation

Translocation involves the shifting of a part of one chromosome to another nonhomologous chromosome. If two non-homologous chromosomes exchange parts, which need not be of the same size, the result is a reciprocal translocation. The reciprocal translocation may be of following types:

1. **Homozygotic translocation** – In homozygotic translocation normal meiosis occur and cannot be detected cytologically. Genetically they are marked by altered linkage group by the fact that a gene with new neighbours may produce a somewhat different effect in its new location (position effect).

2. **Heterozygotic translocation** – In heterozygotic translocation a considerable degree of meiotic irregularity occur.

Genetic Significance of Heterozygotic Translocation

1. The heterozygous translocation produce semi-sterile organisms because between half and two third gametes fail to receive the full complements of genes required for normal development of sex.

2. Some genes, which formerly assorted independently, exhibit linkage relationships after translocation has occurred; a single reciprocal translocation will reduce the number of linkage groups by one.

3. The phenotypic expression of a gene may be modified when it is translocated to a new position in the genome (position effect).

Types of changes in Chromosome Number

The somatic chromosome number of any species, whether diploid or polyploid is designated as **2n**, and the chromosome number of gametes is denoted as **n.**

An individual carrying the gametic chromosome number, *n,* is known as haploid.

A monoploid, on the other hand, has the basic chromosome number, x.

In a diploid species, $n= x$; one x constitutes a genome or chromosome complement.

The different chromosomes of a single genome are distinct from each other in morphology and/or gene content and homology; members of a single genome do not show a tendency of pairing with each other. Thus a diploid species has two, a triploid has 3 and a tetraploid has 4 genomes and so on.

Individuals carrying chromosome numbers other than the diploid (2x, and not 2n) number are known as heteroploids, and the situation is known as heteroploidy.

'The change in chromosome number may involve one or a few chromosomes of the genome; this is known as *aneuploidy.*'

'Heteroploidy that involves one or more complete genomes is known as *euploidy.*'

By definition, therefore, the chromosome numbers of euploids are an exact multiple of the basic chromosome number of the concerned species.

A summary of the terms used to describe Heteroploidy (Variation in chromosome number)

Term	Type of change	Symbol*
HeteroploidA.	A CHANGE FROM 2X	
ANEUPLOID	*One or a few chromosomes extra or missing from 2n*	-2n ± few
Nullisomic	One chromosome pair missing	2n – 2
Monosomic	One chromosome missing	2n – 1
Double monosomic	One chromosome from each of two different chromosome pairs missing	2n – 1 – 1
	One chromosome extra	2n + 1
Trisomic Double trisomic	One chromosome from each of two different chromosome pairs extra	2n + 1 + 1
	One chromosome pair extra	2n + 2
Tetrasomic		
B. EUPLOID	*Number of genomes or copies of a genome more than two*	
Monoploid	One copy of a single genome	x
Haploid	Gametic chromosome complement	n
POLYPLOID	*More than 2 copies of one genome or 2 copies each of 2 or more genomes** Genomes identical with each other*	
1. Autoployploid	Three copies of one genome	3x
Autotriploid	Four copies of one genome	4x
Autotetraploid	Five copies of one genome	5x

Term	Type of change	Symbol*
Autopentaploid	Six copies of one genome	6x
Autohexaploid	Eight copies of one genome	8x
Autooctaploid	*Two or more distinct genomes*	
2. Alloployploid	**(generally each genome has two**	
AABB	**copies)**	
Allotetraploid	Two distinct genomes	(2x1 + 2x2)**
Allohexaploid	Three distinct genomes	(2x1+ 2x2 + 2x3)**
Allooctaploid	Four distinct genomes	(2x1+2x2+2x3+2x4)**

2n = Somatic chromosome number (and complement) and n = gametic chromosome number (and complement) of the species, whether diploid or polyploid.

X = the basic chromosome number (and complement) or genomic number. x1, x2, x3, x4 = Distinct genomes from different species.

** In general, this condition occurs; other situations may also occur.*

POLYPLOIDS – Auto and Allopolyploids

Autopolyploidy

In autopolyploidy are included **triploidy, tetraploidy** and **higher levels of ploidy**.

Autopolyploids are produced directly or indirectly through chromosome doubling, which is briefly considered as follows.

- **Origin and production of Doubled Chromosome Numbers**

   (1) Spontaneous, (2) Due to treatment with physical agents, (3) Regeneration in vitro, (4) Colchicine treatment, and (5) Other chemical agents.

Autopolyploid Crop species — Application in Crop Improvement

Common name	Scientific name	Somatic chromosome number (2n) of the cultivated form	Somatic chromosome numberof related wild species
Potato	*Solanum*	48 (4x)	24 (2x) form of *S.*
Coffee	*tuberosum*	44 (4x)	*tuberosum*
Alfalfa	*Coffea arabica*	32 (4x)	22,66,68
PEANUT	*Medicago sativa*	40 (4x)	14,16,32
Banana	*Arachis hypogaea*	33 (3x)	
	Musa sapientum		22-
Sweet	*(M.paradisiaca)*	90 (6x)	
potato	*Ipomoea batatas*		

Allopolyploidy

Allopolyploids have genomes from two or more species

Several of our crop plants are allopolyploids and as almost always were the creation of new species. Some success has been obtained as is evident from the emergence of **Triticale** as a new crop species in some areas, and the promise shown by some other allopolyploids, e.g., Raphanobrassica and some allopolyploids of forage grasses. Some genera, which contain allopolyploid species, and one or more crop species. The crop species themselves may be allopolyploid or diploid. Genera like **Triticum, Brassica** and **Gossypium** have both diploid and allopolyploid crop spices.

Scientific name	Common name	Gametic chromosome number (n)*	Cultivated/Wild**
A. sativa	Cultivated oats	14	C
B. oleracea	Cabbage	9(C)	C
B. juncea	Rai, Indian mustard	18	C
B. napus	Rape	19	C
Gossypium arboreum	Asiatic (desi) cotton	13 (A2)	C
G. herbaceum	Asiatic cotton	13 (A1)	W
G. thurberi	Wild American cotton	13 (D1)	W
G. barbadense	Sea island (Egyptian)	26 (A2D2)	C
G. hirsutum	cotton	26 (A1D1)	C
Hordeum vulgare	American upland cotton	7	C
Saccharum	Cultivated barley	40	C
officinarum	Noble canes	41,45,46,58,62	C
S. barberi	Indian canes	58,59	C
S. sinense	Indian canes	20-64	W
S. spontaneum	Kans (wild canes)		

Letters within parentheses denote the symbols used for genomes present in the species.

** C = Cultivated; W = Wild.

REFERENCES

Agarwal, P.K. (1993). *Handbook of Seed Testing*. (Eds.). Department of Agriculture and cooperation, Ministry of Agriculture, Govt. of India, New Delhi.

Agarwal, R.L. (2005). *Seed technology*. Oxford & IBH Publishing Co. Pvt. Ltd. New Delhi. Pp. 829.

Agarwal, R.L. (1984). *Ientifiaction of Crop Varieties*. Oxford and IBH Publishing Co., New Delhi.

Allard, R.W. (1960). *Principles of Plant Breeding*, John Wiley & Sons, Inc., New Delhi.

Amar Singh (1982). *Practical Physiology*. Kalyani Publishers. New Delhi.

Anonymous (1928). *Report of Royal Commission on Agriculture*, Abridged Report, Govt. of India, Central Publication Division, Calcutta.

Anonymous (1961). *Agricultural and Horticulatural Seeds: Their Production Control and Distribution*, F.A.O. Agricultureal Studies No. 55, Rome.

Anonymous (1961). *Report on Seed Multiplication Shemes, Committee on Plan Projects, Planning Commission*, Govt. of India, New Delhi.

Anonymous (1971). *Certification Handbook*, Ass. Off. Seed Certificfying Agencies, Publication. No. 23.

Arvind, M.D. and Manuel, L.M. (2002). *Principles and Techniques for Plant Scientist*. Agrobios. India.

Chalam, G.V. Singh, A. and Douglas, J.E. (1967). *Seed Testing Manual*, ICAR and USAID, Publication.

Cowan, J.R. (1972). *Seed Certification*. Seed Biology. Vo. III. Academic Press, New Delhi.

Darnell, J., Lodish, H. and Baltimore, D. (1986). *Molecular Cell Biology*. Scientific American Books Inc., New Delhi.

Dutta, A.C. (2007). *Botany*. Oxford University Press.

Feistrizer, W.P. (1975). The role of seed technology for agricultural development, *Seed Sci. and Technol*. 3: 415-429.

Foster and Twell (1997). *Plant Gene Isolation: Principles and Practices*. John Wiley and Sons, New Delhi.

Freifelder, D. (1990). *Micobial Genetics*. Narosa Publishing House, New Delhi.

Gardner, E.J., Simmons, M.J. and Snutad, D.P. (1991). *Principles of Genetics*. John Willey and Sons Inc. New Delhi.

Gunning, B.E.S. and Steer, M.W. (1996). *Plant Cell Biology*. Johns and Bartlett Publishers, London.

Gupta, P.K. (2000). *Elements of Biotechnology*. Rastogi Publications, Meerut, India.

Gupta, M.L. and Jangir, M.L. (2002). *Cell Biology: Fundamentals and Applications*. Agribios, Jodhpur, India.

Hemantararanjan, A. (ed). (2002). *Advances in Plant Physiology.* Scientific Publishers India Vol. 1. Jodhpur.

Indra, K.V. and Trevor, A.T. (Eds). (2005). *Plant Cell and Tissue Culture.* Springer.

Lincoln Taiz. and Eduardo Zeiger. (2006). *Plant Physiology.* Sinauer Associates. Inc., Publishers, Sunderland, Massachusetts.

Kumar, A. and Purohit, S.S. (2005). *Plant Physiology: Fundamental and Applied.* Agrobios, Inida. P. 776.

Maini, N.S. (1975). *National Seeds Corporation – A Review of Activities,* Seeds & Farms, 1: 6: 11-16.

Malik, C.P. and Srivastava, A.K. (2005). *Textbook of Plant Physiology.* Kalyani Publishers. New Delhi.

Mehta, Y.R., Douglas, J.E. and Singh, A. (1973). *Historical development of seed improvent in India.* Seed Res. 1: 1-12.

Mukherji, S. and Ghosh, A.K. (1996). *Plant Physiology.* New Central Book Agency (P). Ltd. Kolkata, India.

Old, R.W. and Primrose, S.B. (1993). *Principles of Gene Manipulation: An Introduction to Genetic Engineering.* Backwell Scientific Publications, London, UK.

Paliwal, R.L. (1971). *Procedures for maintenance of varieties and production of pure foundation seed stocks and their allocation,* Proc. All India Wheat Research Workers' Workshop.

Paterson, A.H. (1996). *Genome Mapping in Plants.* Academic Press, London.

Pandey, S.N. and Sinha, B.K. (2006). *Plant Physiology.* Vikas Publishing Huse Pvt. Ltd. New Delhi. p. 682.

Salisbrry, B. and Ross, C. (1990). *Plant Physiology.* Prentice Hall of India, New Delhi.

Sheeler, P. and Branchi, D.E. (1987). *Cell and Molecular Biology.* John Willey and Sons Inc., New Delhi.

Shivraj, A. (1987). *An Introduction to Physiology of Cereal Crops.* Oxford and IBH Publishing Co., Pvt. Ltd. New Delhi.

Stent, G. and Calender, R. (1986). *Molecular Genetics.* CBS Publishers, New Delhi.

Vaughan, C.E., Gregg, B.R. and Delouche, J.C. (1968). *Seed Processing and handling,* Handbook No. 1., Miss. State Univ., State College, Miss.

Verma, P.S. and Agarwal, V.K. (1994). *Cytology.* S. Chand and Co., New Delhi.

Walker, J.M. and Gaastra, W. (1983). *Techniques in Molecular Biology.* Croom Helm, London.

World, A. (1980). *The International Sed Testing Association organization, Objects and activities.* Seed Tech News. 10:1: 1-4.